P9-DJL-957

SCARED SICK

Also by Robin Karr-Morse and Meredith S. Wiley

Ghosts from the Nursery: Tracing the Roots of Violence

SCARED SICK

The Role of
Childhood Trauma in
Adult Disease

ROBIN KARR-MORSE

WITH MEREDITH S. WILEY

BASIC BOOKS
A Member of the Perseus Books Group
New York

Copyright © 2012 by Robin Karr-Morse and Meredith S. Wiley

Published by Basic Books,
A Member of the Perseus Books Group

All rights reserved. Printed in the United States of America. No part of this book may
be reproduced in any manner whatsoever without written permission except in the case
of brief quotations embodied in critical articles and reviews. For information, address
Basic Books, 387 Park Avenue South, New York, NY 10016-8810.

Books published by Basic Books are available at special discounts for bulk purchases in
the United States by corporations, institutions, and other organizations. For more
information, please contact the Special Markets Department at the Perseus Books Group,
2300 Chestnut Street, Suite 200, Philadelphia, PA 19103, or call (800) 810-4145, ext. 5000,
or e-mail special.markets@perseusbooks.com.

Designed by Brent Wilcox

Library of Congress Cataloging-in-Publication Data
Karr-Morse, Robin.
   Scared sick : the role of childhood trauma in adult disease / Robin Karr-Morse ;
with Meredith S. Wiley.
      p.   cm.
   Includes bibliographical references and index.
   ISBN 978-0-465-01354-8 (hardcover : alk. paper) — ISBN 978-0-465-02812-2 (ebook)
   1. Psychic trauma.   I. Wiley, Meredith S.   II. Title.
   BF175.5.P75K37 2012
   155.9'3—dc23
                                                                    2011029405

10 9 8 7 6 5 4 3 2 1

For Alice Miller,
whose work showed us that the child
is indeed the father of the man—
and inspires those of us in her shadow
to have the courage to do the same

For we live with those retrievals from childhood that coalesce and echo throughout our lives, the way shattered pieces of glass in a kaleidoscope reappear in new forms and are songlike in their refrains and rhymes, making up a single monologue. We live permanently in the recurrence of our own stories, whatever story we tell.

—MICHAEL ONDAATJE, *Divisadero*

CONTENTS

PREFACE

Confronting challenges to our health, we typically consider potential causes like germs, genetics, diet and environmental toxins. But we often overlook one of the most formative factors of all—the pervasive role of early emotional trauma. Experienced without detection early in life, then held without repair, trauma may lie unseen at the root of many forms of illness that we currently dismiss as genetic or as the inevitable results of aging.

Scared Sick began as our effort to answer a question we have been frequently asked since writing *Ghosts from the Nursery: Tracing the Roots of Violence*. Released just as school shootings erupted across the country, *Ghosts* drew upon emerging science to explain how child abuse and neglect can alter the brain, paving the way for aggression and violence. But many readers wondered: What happens to the majority of abused and neglected children who don't become violent? Do they actually emerge unscathed?

What we found is surprising. While it is true that most abused or neglected children do not become violent, chronic early trauma exacts an enormous price not only in emotional but in physical and behavioral outcomes. And child abuse and neglect are only the tip of the iceberg. Early emotional trauma is also a common by-product of many routine practices unrecognized as traumatic, experiences that range across cultures, religions, ethnicities, race and income levels.

The equation described in the following pages in fact applies to all of us—our lives, our families, our futures—and those of everyone we know. The degree to which emotionally traumatic experiences pave the way for disease is a unique calculation for each individual, mediated by several factors, including genetics, timing, the intensity and frequency of trauma, and the presence or absence of repair. Because most chronic disease builds

slowly and does not manifest until later, diagnoses are typically disconnected from their early developmental roots.

Before we begin, a few qualifiers are in order. First, it is clear that
when we get sick there are often many factors at work beyond emotional
ones. Genes and germs, injuries and aging play obvious roles. Second,
we know that fear and trauma are inherent in the human condition. The
point of *Scared Sick* is that with heightened awareness we can greatly reduce the rate of emotional trauma in our youngest children, when fear
renders developing nervous systems relatively more vulnerable to later
trauma and disease. Finally, we recognize that many of us have experienced trauma. For those of us far from childhood, *Scared Sick* also shows
us that there is much we can do at any age to heal and reduce the cumulative toll of chronic fear and trauma on our health.

ACKNOWLEDGMENTS

Scared Sick is conceived and written with deep gratitude to the thinkers who have paved the way for all of us to comprehend and heal trauma in our lives: Dr. Bessel VanderKolk, Dr. Bruce Perry, Dr. Allan Schore, Dr. Daniel Siegel, Dr. Bruce McEwen and Dr. Robert Anda; and especially to Dr. Vincent Felitti and Dr. Robert Scaer, who helped shape this manuscript through long hours of conversation and explanation of their pioneering work.

Without the contributions of several additional people, this book would not have come to be. Including:

- Wanda Kaczynski, who spent hours sharing her memories of her son, Ted. Her wisdom and courage in the face of her family's tragedy is inspiring. As is her younger son, David, who not only facilitated the interviews with his mother but has devoted his life to the elimination of the death penalty and is an unwavering and powerful advocate for the rights of crime victims.
- Jordan Karr-Morse, who volunteered his professional skills and considerable talent as a filmmaker to film Wanda Kaczynski and to produce a book trailer based on the interview to help get the word out.
- Arielle Bernstein, who juggled demanding jobs, graduate school and personal commitments to provide us with whatever we were looking for and who kept a highly informed eye on all the details.
- Mitch Douglas, our agent and friend, who has been there for us all along.
- Colin Karr-Morse, who once again survived the brunt of the daily demands of producing a book and supported us every step of the way.

Also, a thank-you to David Kass, Miriam Rollin, Jeff Kirsch and Amy Taggert-Dawson, the senior management team at Fight Crime: Invest in Kids, who went out of their way to allow Meredith to juggle her job requirements to spend blocks of time working with Robin.

INTRODUCTION

Fear has shadowed human life since our species emerged. But the velocity of change in our current environment and the nature of the fears we face as a consequence have evolved much faster than our biological systems for dealing with them. We are currently witnessing a fracturing and amalgamation of cultures unprecedented in human history. And as technology shrinks and transforms our world, the advances we have made that enabled us to defeat many kinds of physical challenges have in turn created complex threats of their own. Among these is our failure to recognize and protect our elegant, intrinsic systems for perceiving and responding to threat.

On a societal level, Americans regularly wake up, work, parent, drive, play, eat and sleep with the twin offspring of fear—anxiety and depression—holding court in their brains and bodies. This is our shared daily bath: in our homes, on the road, in the workplace. The result? Soaring rates of addiction, anxiety, depression, attention disorders and post-traumatic stress. And an epidemic of diabetes, obesity, heart disease and inflammation-driven conditions like arthritis signals that something is very out of balance in our systems.

Not only physicians and traumatologists but sociologists are pointing out parallels between growing rates of individual and societal trauma and impulsive, often irrational decision-making, including overconsumption. A 2010 article in the *New York Times Magazine* by Judith Warner, titled "Dysregulation Nation," explores the lack of systemic regulation that has presaged major disasters affecting us in the past several years. Warner observes that the oil fiasco in the Gulf of Mexico in April 2010 was only the latest example of the dysfunction of key regulatory systems. Scrutinizing the 2008 banking meltdown, the collapse of the housing market, and the failure of the levees in New Orleans after Hurricane Katrina, Warner illuminates similarities between the regulatory dysfunction of large systems and the lack of self-regulation within individuals, including "appetite, emotion, impulse and cupidity," which, she argues, may well be the defining social pathology of our time: "The signs that something is amiss in our inner mechanisms of control and restraint are everywhere."[1]

Nowhere is this dysfunction more apparent than in our health. Consider the following snapshot of American health from the Centers for Disease Control and Prevention (CDC) in the spring of 2010:

- We have one of the shortest life expectancies of any industrialized nation, lagging behind more than forty other countries, including most of Europe and Japan.[2]
- Nearly half of all Americans have high blood pressure, high cholesterol or diabetes. Many have more than one of these conditions: one in eight Americans has at least two, and one in thirty-three has all three.[3]
- Nearly one-third of us have high blood pressure.[4]
- One-third of all deaths in our nation are due to heart disease or stroke.[5]
- More than one-third of American adults are obese. In addition, 68 percent of adults and one-third of all children and teens in this country are overweight.[6]
- From 1980 to 2009, the number of Americans with diabetes more than quadrupled; nearly 10 percent of the U.S. population now has diabetes.[7]
- Twenty-six percent of adults over eighteen suffer from a diagnosable mental disorder.[8]
- Eighteen percent of adults in the United States over age eighteen suffer from an anxiety disorder.[9]
- Nearly one-tenth of American adults (19.4 million) meet the clinical criteria for a substance abuse disorder—either alcohol or drugs or both.[10] One in five has a spouse, sibling or child who has been addicted at some point to alcohol or drugs.[11]

SUFFER THE LITTLE CHILDREN

For American children, the handwriting is on the wall:

- Among the seven largest industrialized nations in the world, the United States ranks last on infant mortality rates and longevity.[12]
- The well-being of American children ranks twentieth among twenty-one rich democracies, behind Poland, Greece and Hungary.[13]
- One in three children born five years ago in this country will develop diabetes in their lifetime.[14]
- Child abuse death rates in the United States are far higher than in all of the seven largest industrialized countries: three times higher than in Canada and eleven times higher than in Italy.[15]
- Just under five U.S. children die every day as the result of child abuse. Three out of four are under four years of age; nearly 90 percent of the perpetrators are the biological parents.[16]
- American children are the new frontier for sales of prescription drugs. One-quarter of the children insured by Medco, a large manager of drug benefits, took prescription medicine to treat a chronic condition last year, including asthma, attention deficit/hyperactivity disorder (ADHD), obesity, high cholesterol, heartburn and diabetes—representing a spending increase for drug costs of 10.8 percent, more than triple the amount for seniors.[17]
- In 2005, 15.5 percent of all babies born in the United States were low-birthweight and/or preterm at delivery.[18]
- Just over 20 percent (one in five children) either currently or at some point have had a seriously debilitating mental disorder. Thirteen percent of eight- to fifteen-year-olds have had a diagnosable mental disorder within the past year.[19]
- An estimated 26 percent of all children in the United States will experience or witness a traumatic event prior to age four.[20]
- One in one hundred infants each year is born with fetal alcohol spectrum disorder (FASD), the leading known preventable cause of mental retardation and birth defects in the Western world and a leading known cause of learning disabilities. More children have

FASD than are affected by autism, Down syndrome, cerebral palsy, cystic fibrosis, spina bifida and sudden infant death syndrome (SIDS) combined.[21]

• Of children ages three to seventeen, 4.7 million (10 percent of boys and 8 percent of girls) have a learning disability.[22]

Scared Sick is the story of the connections between fear and ill health. No longer sidelining this story to the domain of metaphysics, medical researchers across the world are unveiling in biological terms how it is that our experiences affect our biology, particularly when these experiences are chronic, happen early in life, and remain unrecognized. You will discover the role of early emotions in shaping the organization of the central nervous, endocrine and immune systems and the physical mechanisms that render children particularly vulnerable to the effects of fear and trauma. Precisely because the youngest children have no experiential ballast against these forces, early chronic fear can have a formative role in lifelong health. You will comprehend how early fear triggers disease by dysregulating the HPA axis, activating the vagus nerve, and catalyzing epigenetic mechanisms that facilitate the expression of genetic disease. You will discover how this equation is individually tailored by the balance between protective factors and risk factors throughout development, particularly by strong positive emotional connections in early development. And you will find potential routes to heal emotional trauma—for yourself or for someone you love.

We will introduce you to the research in the order we discovered it so that you can connect the dots in the same sequence we did. We start with groundbreaking research by Kaiser Permanente that suddenly and inexorably has cast doubt on current assumptions about traditional explanations for many forms of disease. We explore how we came to question the "old news"—everything we have historically believed to be true regarding health and disease. Then we look at the "bad news," that is, what new imaging technology is now revealing about the biology of stress and trauma and how our experiences shape our biology, potentially triggering not only being "scared sick" but being "scared stiff"—to the point of death (Chapters 2 and 3). Next we examine the overlooked role of chronic fear in daily situations affecting fetal, infant and toddler lives and specific diseases in adulthood that correlate with early experiences of chronic stress and trauma (Chapters 4, 5 and 6). We then turn to the "new news" about genetics in Chapter 7, focusing particularly on the emerging science of "epigenetics"—with amazing information for us all. Finally, we arrive at the "good news," beginning with the powerful role of attachment relationships (Chapter 8) and new directions for adults seeking to heal emotional trauma (Chapter 9).

Whatever our own history, there is much we can do to change these outcomes and to intervene—even in our last decades—in our emotional and physical health, as well as our children's. The final chapter of *Scared Sick* explores the huge implications of the research for our daily lives and for altering insensitive practices in societal systems to support health and prevent disease from the beginning of life. There may be no more crucial way to make a difference than this.

CHAPTER 1

Monster in the Closet

Trauma in the Body

IT IS ALMOST IMPOSSIBLE to pick up a magazine, thumb through a newspaper, or turn on a television news program without being bombarded by recent research on obesity, the latest news on dieting, or the newest celebrities lending their faces to our nation's problem with addiction. The faces change and the advice varies, but the message remains the same: we Americans have a problem with overindulgence in habits that hurt us. "Behavioral health" is a hot topic because these issues permeate every segment of society and impact each of us directly or indirectly. We tend to write off obesity and addiction as either genetically determined or as inescapable aspects of contemporary living. But the closer these concerns come to affecting our own daily lives, the less cavalier we become. It's hard to be dismissive about our own diminishing health or that of someone we love in the face of a serious diagnosis, especially when we learn that the diagnosis stems directly from weight gain or addiction.

But what if neither addiction nor obesity is inevitable? What if these conditions arise, as often as not, from a mostly preventable blindness about what humans need to develop constructively?

The first indications of this reality have long been in front of us. The addictions—our hunger to "fill the void" with food or alcohol or drugs, gambling or shopping—were the first symptoms of behavioral ill health

1

that led unsuspecting researchers to the significance of early experience. The single most common behavioral health issue, and one that many of us stumble over, is obesity. Oprah Winfrey has been the face of this issue for millions. For years she has battled her weight, in spite of her access to the best diets, personal trainers and exercise regimens. Having hit the "dreaded 200," Winfrey revealed in late 2009 that her exquisitely conceived costume for President Barack Obama's inauguration would fade into history unworn—because it didn't fit her ballooning (still voluptuous) figure. Embarrassed and exhausted, Oprah said: "It's not about the food. It's about using food. Abusing food. Too much work. Not enough play. Not enough time to come down. Not enough time to really relax. I am hungry for balance."[1] Mirroring the thoughts of millions of women in this nation, Oprah admitted, "So here I stand forty pounds heavier than I was in 2006. . . . I'm mad at myself. I'm embarrassed. I can't believe that after all these years I'm still talking about my weight. I look at my thinner self and I think, how did I let this happen again?"[2]

Weight loss is a huge market. Many factors are commonly blamed for the increasing number of fat Americans: fast foods; loss of time to prepare fresh foods, whose relatively higher cost may lead to increasing reliance on prepared food; too much TV; too little exercise; the increasing idleness of American schoolchildren. Genetics and organ diseases such as hypothyroidism can also play a role. But they don't explain all or even a majority of instances of obesity. The sheer number of diets and weight-loss products and the ineffectiveness of most of them indicate that we're missing something important. Oprah, being Oprah—a woman whose intelligence has enabled her to grow out of an impoverished, chaotic and abusive childhood into the incredible force she has become as an adult—intuitively acknowledges: "My greatest failure was in believing that the weight issue was just about weight. It's not. It's about sexual abuse. It's about all the things that cause people to become alcoholics and drug addicts."[3]

The health consequences of being overweight are wide-reaching and the topic of much alarm in our nation. Simply being overweight (body mass index [BMI] of 25 to 29.9) or obese (BMI over 30) significantly increases the risk of disease, including hypertension, type 2 diabetes, coronary heart disease, stroke, gallbladder disease, osteoarthritis and some cancers, such as breast and colon. Approximately forty different diseases are linked to obesity. The ramifications are staggering. Thirty-four percent of American adults are obese. In addition, 68 percent of adults and one-third of all children and teens in this country (25 million kids) are considered obese or overweight.[4] Roland Sturm, a senior economist with the Rand Corporation who has overseen several studies on obesity's impact on the quality of life, says: "An obese 30 year old has as many chronic conditions as a normal weight 50 year old and reports quality of life that is worse than a 50 year old." Asked what he thinks we should do, Sturm suggests, "Maybe we should start by trying to create an environment that prevents obesity in the first place, especially for children."[5]

The rise in obesity has been accompanied by an explosion in diabetes, which more than quadrupled between 1980 and 2009, from 5.6 million cases to 24 million. The incidence of type 2 diabetes in children is skyrocketing as well.[6] The Centers for Disease Control and Prevention (CDC) has predicted that one in three children born since 2003 in this country will develop diabetes. Diabetics are two to four times more likely to develop heart disease or have a stroke and three times more likely to die of complications from flu or pneumonia. Also related to obesity is a stunning rise in the major precursors to cardiovascular diseases. In the spring of 2010, the CDC released findings that nearly half of all Americans have high blood pressure, high cholesterol or diabetes.[7]

Osteoarthritis is also on the rise, owing to fat. Every extra pound puts additional weight on joints never intended to carry the extra load. According to the CDC, 51.2 million Americans suffer from osteoarthritis, and by 2030, as baby boomers age, the number of cases will increase by over 40 percent.[8]

There is little question that Americans are losing the battle of the bulge. Certainly it isn't for lack of trying: the commercial weight-loss market is huge. In spite of a plethora of books and theories on how to lose weight quickly and with minimal discomfort, confusion reigns supreme. Millions of Americans are on a diet on any given day; few are succeeding. Obese teenagers are a new and growing market for stomach bands. More than a thousand American teenagers underwent bariatric surgery in 2007. According to leading specialists, including Reginald Washington, a Denver pediatric cardiologist and past cochairman of the

American Medical Association's childhood obesity task force, "this is not an end to treatment, it's a way to get started."[9] A desperate measure perhaps, but the reality, according to Washington, is that traditional diet and exercise are effective only 30 percent of the time.

Dr. Emma Patterson, medical director of the Obesity Institute at Oregon's Legacy Research Project, echoes the concerns of a growing group of surgeons across the country when she says: "Why wait until they have high cholesterol, kidney and heart problems or need a new knee?"[10] Clearly, advertising campaigns targeting kids, the ready availability of soft drinks and junk food; lack of time, information, and resources to buy fresh unprocessed foods; and the lack of safety in some impoverished neighborhoods have all played a role. In fact, teenagers living in poverty are 50 percent more likely than their wealthier peers to be overweight. We know that poverty and poor neighborhoods carry additional loads of stress for families, which raises an obvious question: is there a deeper force at work?

Dr. Sonia Lupien, on staff at the Douglas Hospital in Montreal, believes there is, commenting to a *Globe and Mail* reporter, "I think stress is a major factor in this."[11] Unquestionably, social status has a significant indirect impact on health, leading to higher rates of cardiovascular disease, diabetes and other lifestyle-related diseases that result, in turn, in an overall lower life expectancy. Low-income people not only eat a poorer diet, they also experience more stress and have fewer resources that support healthy interventions, such as health insurance. Lupien's research, focused on 450 children from low- to high-income families, found "a three-fold increase in stress hormones in low-income children compared to rich children."[12]

Is it possible that we are blind to a deeper force at work? Physicians, researchers, addiction experts and individuals like Patty Worrells are increasingly voicing this question. Patty's story appeared in an unlikely publication for such coverage: the *Wall Street Journal*. Patty was a food addict whose 3:00 a.m. binges on half-gallon tubs of ice cream and eight cinnamon rolls for breakfast finally caught up with her.[13] In her midforties, Patty, at five-foot-four, weighed 265 pounds, was diabetic and arthritic. She had gastric bypass surgery and lost 134 pounds in a year. Patty was elated with the changes in her body, which enabled her to fit comfortably into car and airplane seats, restaurant booths and crowded shopping aisles. Then her life shifted again. Eighteen months after the surgery, her food cravings were gone. But she found herself downing fifteen to twenty shots of tequila almost every night. Normally soft-spoken, Patty developed a reputation for wild partying, often waking with scratches and bruises she couldn't account for. Patty's domestic partner said, "She became a monster!"

Patty says: "I knew I was going to die." She recognized the pattern. Her father had died from alcohol abuse, and her sister struggled with addiction. In an effort to protect her mother from confronting yet another alcoholic in the family, Patty struggled for months to hide her drinking. Her turnaround began when her partner, exhausted one evening by Patty's rages, picked up the telephone and dialed Patty's mother. "Listen to your daughter," she said and positioned the receiver to pick up Patty's alcohol-fueled cursing. The next day Patty, horrified that her mother now knew, attended her first twelve-step meeting. Three weeks later Patty's mother called with devastating news: Patty's sister had died from an overdose of Xanax. Patty recalls being grateful she was sober as she drove to be at her mother's side.

At the time the article was written, Patty was committed to recovery, attending four twelve-step meetings a week, but her course was riddled with relapses. Patty is far from alone in her path from food addiction to bariatric surgery to another addiction. Treatment centers like the Betty Ford Center in Rancho Mirage, California, report increasing numbers of bariatric surgery patients seeking help with new addictions. And alcohol abuse is a hot topic for online support sites like the Weight Loss Surgery Center, which has more than ten thousand subscribers on its e-newsletter list.[14] Gastric bypass patients are extremely vulnerable to alcohol abuse, with estimates ranging from 5 percent to 30 percent of patients switching from food to alcohol addiction after the surgery. Numbers for those who switch from eating to smoking, according to estimates, are not far behind.[15]

Replacing one addiction with another—known as "addiction transfer"—is not limited to switches from food to alcohol. The Betty Ford Center notes that about one-quarter of relapsed alcoholics replace alcohol with another type of drug, often opiates.[16] There is a growing understanding among researchers that the foundation for addiction transfer is biological. But where it once would have been attributed to genetics, researchers are now looking at brain chemistry. "There are similarities in the circuits that are affected in the processes of addiction and obesity," says Dr. Nora Volkow, director of the National Institute on Drug Abuse.[17] Of particular interest is the observation that obese people, alcoholics and drug addicts all have below-normal levels of dopamine—a hormone associated with pleasure—which contributes to cravings. Addictive substances temporarily boost dopamine.

Research into the pathways of addiction and its underlying brain chemistry has exploded. Dozens of clinical trials are under way at the National Institutes of Health (NIH), some focusing on developing new drugs, others on following the effects of medications previously used for other diseases. Toprimate, a drug used to treat epilepsy and marketed under the name Toprimax, is now being studied for alcohol and cocaine addiction, binge-eating and compulsive gambling. And the antidepressant bupropion (Welbutrin) is often being prescribed for withdrawal from alcohol as well as to treat gambling, obesity and nicotine dependence. The search continues for a pill that will curb our cravings, dampen our appetites, and boost our mood.

But a growing group of researchers are also looking more deeply at what lies beneath this faulty brain chemistry. "Why," they are asking, "are dopamine levels so low in some people? Are there common variables at the root of this reality?" Dr. Gabor Maté of Vancouver, Canada, author of several best-selling books, the most recent of which is about addiction (*In the Realm of Hungry Ghosts*), says:

> If you look at the brain circuits involved in addiction . . . we're looking for endorphins in our brains. Endorphins are the brain's feel good, reward, pleasure and pain relief chemicals. They also happen to be the love chemicals that connect us to the universe and to one another. Now, that circuitry in addicts doesn't function very well. . . . The issue is, why do these circuits not work so well in some people, because the drugs in themselves are not surprisingly addictive. And what I mean by that is, is that most people who try most drugs never become addicted to them. And so, there has to be susceptibility there. And the susceptible people are the ones with these impaired brain circuits, and the impairment is caused by early adversity, rather than by genetics.[18]

This concept was not yet on the horizon in the 1980s when Kaiser Permanente in San Diego created a clinic for chronically obese patients. The initial effort, headed by Dr. Vincent Felitti, grew out of a very advanced department of preventive medicine that he had created for Kaiser patients. Felitti's vision was—and is—to return family practice and internal medicine to their roots in preventive care, as opposed to operating as a symptom-driven response to illness. This department became the largest single-site medical evaluation facility in the world, serving 58,000 people a year. When we met with Felitti in California in 2005, he recalled:

> In the course of doing that work it became quickly evident that we needed to create our own risk abatement program. The first project we put together was a weight program. . . . We happened to be in possession of a very powerful technology that allowed us to safely take people's weight down about three hundred pounds a year. But after about five years of doing this, we realized we had a very big problem, namely, an enormously high dropout rate that was almost exclusively limited to people who were successfully losing weight . . . which drove me nuts! It was wildly counterintuitive, and it was ruining the reputation of the program. . . . It was the exploration of that concern that led us into the ACE Study.[19]

Puzzled by the high dropout rate among patients who were successfully losing weight, Felitti decided to interview them in depth about their lives and the reasons they chose to quit the program. He interviewed 186 patients, and then, stunned by his findings, he had five colleagues interview an additional 100. Their unexpected results provided strong evidence that obesity, which the doctors viewed as the problem, was actually a solution to deeper problems. These patients overate to assuage their feelings and used fat to buffer an underlying need not to be vulnerable—physically, emotionally or sexually. Obesity was key to their sense of self-protection. One patient, a woman who had gained 105 pounds in the year after she was raped, said simply: "Overweight is overlooked, and that's the way I need to be."[20] Typically the patients had never discussed these issues with anyone, not even their physicians. Felitti observed, "We found the simultaneous presence of opposing forces to be common;

many of our weight program patients were driving with one foot on the brakes and one on the gas, wanting to lose weight but fearful of change."

As Felitti was working with obesity in San Diego, Dr. Robert Anda, an epidemiologist and CDC researcher in Atlanta, was simultaneously studying the psychosocial origins of physical diseases, obesity and risk-taking behaviors such as drug and alcohol abuse, smoking and risky sex. Following a presentation Felitti gave in Atlanta on his study of weight program dropouts, he met with Anda, who recognized the significance of these findings. The two physicians ultimately came together to develop the Adverse Childhood Experiences (ACE) Study. The average age of the Kaiser patients was fifty-seven. They were middle-class and preponderantly white, just over 50 percent were female, and 54 percent had a college degree. The ACE questionnaires were mailed to patients two weeks after they had been evaluated at Kaiser's Health Appraisal Center.

Felitti explained that there are really only three major sources of information in medicine: the medical history, the physical exam and laboratory studies. Western doctors and patients tend to focus on lab studies. Yet, he says, "for more than a century, knowledgeable physicians have all concluded that about 80 percent of the time diagnosis comes out of history. Not out of MRIs, not out of physical exams, but out of history. So we put together a rather remarkable questionnaire to gather history."[21]

The ACE questionnaire included questions that were pertinent to child maltreatment, including recurrent physical, emotional or sexual abuse. Five other categories asked about growing up in a household characterized by dysfunction due to: (1) alcoholism or drug abuse; (2) an incarcerated household member; (3) someone who was chronically depressed, mentally ill or suicidal; (4) the mother being treated violently; or (5) parents being separated, divorced or otherwise lost to the patient in childhood. The questions were not ephemeral; they were designed to elicit information that could be objectively evaluated. To minimize the degree of subjectivity involved, the questions pinpointed the number, frequency and intensity of specific behaviors in the past lives of patients. The ACE score measured categories of adverse experience, not the number of incidents. Thus, an individual molested repeatedly by several persons got one point for the category. A patient who reported none of the experiences in the ACE categories had a score of 0. A patient who responded yes to one or two categories had a score of 1 or 2, and so on. The researchers then cross-referenced the current health status of more than seventeen thousand adults with their responses to the ACE questionnaire. They have now followed the same group for fifteen years to assess the relationships between their ACE scores and pharmacy costs, doctor office visits, emergency room use, hospitalization and death.

The study looked at the correlations between adverse childhood experiences and ten risk factors associated with the leading causes of morbidity and mortality in the United States, including smoking, severe obesity, physical inactivity, depressed mood, suicide attempts, alcoholism, drug abuse, injection drug abuse, a high lifetime number of sexual partners and a history of having a sexually transmitted disease. The researchers also examined the relationship between early adverse experiences and the diseases that are among the leading causes of death in the United States: ischemic heart disease, cancer, stroke, chronic bronchitis or emphysema, diabetes, skeletal fracture, liver disease and hepatitis or jaundice.

The single most stunning finding in the ACE Study is the sheer prevalence of adverse childhood exposures. Two-thirds of respondents reported experiences in one or more of the categories. More than a quarter of them had grown up in a household in which there was an alcoholic or drug abuser; the same percentage said that they had been beaten as children. Forty-two percent were exposed to two or more categories of abusive experiences, and one in nine were exposed to five or more. Felitti says the data show that a person exposed to one category of abusive experience has an 87 percent chance of exposure to at least one other category, and a 50 percent chance of exposure to three or more. These experiences tend to occur in clusters. For example, a child in an alcoholic home is typically exposed to other kinds of abuse; no one grows up in a household where Mom is beaten or Dad is in prison but everything else is fine. ACE Study researchers also noted what is called a "dose-response" effect: the higher the ACE score, the worse the outcomes.

Among the findings was that the likelihood of smoking increases with each point on the ACE scale. With a midrange ACE score of 4, a person is twice as likely to smoke, four times more likely to suffer from emphysema or chronic bronchitis, and two and a half times more likely to have chronic obstructive pulmonary disease than a person with an ACE score of 0. Compared with a person with no history of adverse childhood experiences, a person with an ACE score of 4 or higher is seven times more likely to be alcoholic, and six times more likely to have had sex before age fifteen. This person is also twice as likely to have heart disease and twice as likely to have cancer. In addition, those with an ACE score of 4 or higher are forty-six times more likely to be depressed, and twelve times more likely to commit suicide than a person with an ACE score of 0.

Being overweight was actually one of the outcomes less highly correlated with early maltreatment—compared to the extraordinary correlations with depression, suicide and drug abuse. But subsequent research, including a 2009 study reported in the journal *Obesity*, is now pointing to a link between child abuse and obesity.[22] Based on court records from 1967 to 1971, researchers compared the adult body mass indexes of 410 children who were substantiated victims of physical abuse, sexual abuse and neglect with 303 similar children who had not been mistreated. Adults who had been sexually or physically abused and neglected as children had significantly higher BMI scores than those who had not been abused and neglected.[23]

As of 2009, the ACE data also have begun to link childhood trauma to premature death. Using the National Death Index, researchers identified 1,539 deaths in the ACE group between 1995 and 2006. They found that people with an ACE score of 6 or higher on average died nearly twenty years earlier than those with a score of 0 (60.6 years versus 79.1 years).[24] Although the researchers cautioned that this initial look at the link between mortality and childhood trauma involved a relatively small sample, they deemed it statistically significant. High-scoring ACE Study participants were dying substantially younger, "even if you take the absolute number out of it," noted Dr. David Brown, an epidemiologist at CDC.[25]

So how are adverse experiences in childhood linked to health risk behaviors and adult disease? The ACE Study reveals two paths. One involves the use of coping substances like nicotine, drugs or alcohol as a contributing factor in the development of disease. But high-risk behaviors don't explain it all. The second route from adverse childhood experiences to disease is in fact more direct. Participants with an ACE score of 7 who had no evidence of risk-taking behaviors in their history nevertheless had a 30 to 70 percent higher risk of ischemic heart disease in adulthood. Those who had a score of 4 or higher had two to four times the rates of anger and depression, and two to four times the rates of hypertension and diabetes as those with a lower score. As the number of ACE experiences increased, so did the chances of the individual experiencing cancer, chronic lung disease, skeletal fractures and liver disease.

The correlations between addictions—nicotine, alcohol and illicit drugs—and early adverse experiences were so strong that the researchers concluded that "addiction" is more attributable to characteristics intrinsic to early life experiences than to characteristics within the drugs themselves. Felitti believes that drug use is a form of self-medication, an attempt to deal with problems that are well concealed by "social niceties and convention." Citing the fact that a boy with an ACE score of 6 has a 4,600 percent increase in the likelihood of abusing intravenous drugs later, Felitti said:

> Is drug abuse self-destructive or is it a desperate attempt at self-healing? This is an important question because if the answer is self-healing, primary prevention is far more difficult than anticipated—possibly because incomplete understanding of the benefits of so-called health risk behaviors causes these behaviors to be viewed as irrational acts that have only negative consequences. Does this incomplete view of drug abuse leave us mouthing cautionary platitudes instead of understanding the cause of our intractable health problems?[26]

Felitti, Anda and colleagues minced no words about the connections substantiated by their research: "Early childhood trauma can lead to an array of negative health outcomes and behaviors, including substance abuse, among both adolescents and adults. . . . The effects of adverse childhood experiences transcend secular changes such as increased availability of drugs, social attitudes toward drugs, and recent massive expenditures and public information campaigns to prevent drug use."[27]

Compounding Felitti's concern is the linkage between addiction and suicide. Felitti, Anda and their team concluded: "A powerful graded relationship exists between adverse childhood experiences and risk of attempted suicide throughout the life span. Alcoholism, depressed affect and illicit drug use, which are strongly associated with such experiences, appear to partially mediate this relationship."[28]

"Unfortunately, these problems are both painful to recognize and difficult to cope with," Felitti says. "Most physicians would far rather deal with
traditional organic disease. Certainly, it is easier to do so, but that approach
also leads to troubling treatment failure and to the frustration of expensive
diagnostic quandaries where everything is ruled out but nothing is ruled
in. We have limited ourselves to the smallest part of the problem—the part
where we are comfortable as mere prescribers of medication. Which diagnostic choice shall we make? Who shall make it? And if not now, when?"[29]

In summary, adverse experiences in childhood generate strong emotions in children. The feelings and the chemistry they generate become
the "monsters" in our "closets," the closets a metaphor for our bodies,
our physical selves. As with the childhood notion of a "monster in the
closet," the invisibility of the impact of emotional trauma on the body
only makes it more powerful.

Unrecognized, the now-silenced cry of the child takes the form of
physical dysregulation in key systems that regulate health. These internal
changes in turn line the path to risk-taking behaviors, including drugs, alcohol, nicotine, addictive eating and risky sex—choices that surface in
preadolescence, adolescence or adulthood. Clearly an effort to cope, these
self-soothing efforts are ultimately counterproductive when they multiply
health risks exponentially. Separately or in combination, addiction and
the cumulative effects of a stress response system operating out of balance for too long can catalyze genetically shaped health problems that
are typically not diagnosed until late adolescence or adulthood.

Television shows like *The Biggest Loser* and *Celebrity Rehab* focus on
the effects of an overlooked factor: frozen fear, the result of cumulative
emotional trauma. Although we recognize that many children who experience early emotional trauma, violence or neglect go on to fill our
prison cells as adults, we have believed that the majority somehow emerge
unscathed; in referring to such individuals as "resilient," we overlook the
high price they are paying in terms of their physical well-being, to say
nothing of the cost to their mental health. We have attributed their illnesses primarily to causes beyond our reach, the result of genes or the inevitable legacy of aging. In Felitti's words: "The attribution to genetics is
almost embarrassingly blatant as an escapist device from recognizing one's
own emotions and hence those of other people."[30]

Epidemic levels of obesity and addiction are what Felitti has called "an unconscious solution to unrecognized problems dating back to childhood." Felitti and Anda conclude that all of the adverse childhood experiences they measured, while often well concealed, are unexpectedly common in our mainstream, middle-class population and have a profound effect on adult health a half-century later.[31] Felitti summarized the impact on the social fabric of our nation:

> So essentially we stumbled into the major engine underlying the most common public health problems in the country. . . . In essence, what we saw was that huge amounts of what you see in adults coming through internal medicine is really the result of what was present but unseen in pediatrics . . . and not biomedical disease, which I once would have thought. The question is: how might something that happens to little kids affect their health fifty years later? . . . As a young doctor, I would have proposed maybe you get rheumatic fever as a kid and fifty years later, if you're still alive, you get heart disease. . . . That's all true. But the real action lies in the translation of destructive life experiences in childhood slowly into biomedical disease decades later. And the whole link is lost because of time, because of the shameful nature of this, because of secrecy, and because of social taboos. Nice people don't talk about this stuff—especially not doctors![32]

While Felitti, Anda and their colleagues have yet to break down the ACE data by age at the point of exposure (hard to do since most of us have little conscious memory of our first three years of life), Felitti concluded:

> Though we didn't pursue this specific point in the ACE Study, it's overwhelmingly clear from individual interviews that "early" makes a greater difference because there is less potential then for having had good experiences before some disruptive force is set in place in the child's self-regulatory system. If something terrible happens to you as a teenager, at least there's a possibility that good things will have happened earlier in your life which could offset the trauma. At least the possibility is there.[33]

Anda concurred: "I personally believe that the ACE score is a surrogate for prenatal, maybe even antenatal experience—prenatal experience through the time that people have autobiographical memory when brain development is most important and explosive."[34]

The research on an extensive list of diseases traditionally thought to be genetically driven is now being reconsidered in the light of the ACE findings, which point to a new understanding of these diseases as the result of a rich interaction among genetics, physiology and experience.

Take a look at a partial list of diseases that researchers suspect are affected by negative emotions in response to stress or trauma:

Type 2 diabetes
Crohn's disease
Alzheimer's disease
Hypertension
Irritable bowel syndrome
Cardiovascular disease
Morbid obesity
Osteoarthritis
Anxiety and depression
Fibromyalgia
Chronic fatigue syndrome
Chronic pain syndrome
Addiction to drugs, alcohol and nicotine
Cushing's syndrome
Anorexia nervosa
Osteoporosis
Ulcerative colitis
Susceptibility to forms of cancer (including breast and melanoma)[35]

BUT WHAT ABOUT GENES?

"But aren't genes the culprits behind heart disease, hypertension, high cholesterol, diabetes? My dad had the same medical profile I do. Surely this health conversation is mostly about genes."

When at middle age we come up with a disease that a parent has had, we tend to assume a genetic explanation. For example, we often think that many forms of cancer are genetically linked. And many probably are. But cancer is seldom a case of "nature" acting alone. If diseases like hypertension, cancer or schizophrenia are all about DNA, what explains differing susceptibilities to these diseases between identical twins with identical DNA? Our view of genes as staid and immutable is shifting. Do some conditions run in families? Certainly. Some diseases are entirely genetically driven; even the most stable and loving family, for instance, will not be able to prevent a dread disease like Fanconi's anemia. But a surprising number of diseases are "all in the family," not solely because family members share genes, but because the family literally embodies the fear-filled experiences its members have also shared. In the next chapter, we take a look at the biology behind being "scared sick."

CHAPTER 2

Things That Go
Bump in the Night

The Biology of Stress and Trauma

FOR MOST OF US, fear is a double-edged sword. Fueling forms of entertainment ranging from reality television to media coverage of daily news, to race-car driving, to amusement park rides, violent movies and video games, fear induction is a hot commodity—at least in measured doses. Instinctively, we are captured, challenged, intrigued, motivated—and sometimes paralyzed—by this primal emotion. While essential to our survival, fear can also—if overstimulated—become the unintended emissary of death. It's not surprising, then, that many of us have a love-hate relationship with fear. Throughout history, political systems have harnessed and manipulated this aspect of human nature at least as effectively as today's producers of reality television.

Under less threatening and more controlled circumstances, fear provides a source of thrills, energy and enhanced learning. Most of us can easily remember how fear of losing the game or failing the test delivers the rush that fuels the win or the high score. Fear can be exhilarating, as in extreme sports like mountain climbing and surfing giant waves. It can also crystallize moments in memory, especially those recorded with deeply engraved emotions, some of which we wish we could erase. In our earliest lives, even before we mastered words or reason, we may have stored adrenaline-etched memories in our primitive brains from a single searing experience, such as avoiding a snarling dog or pulling little fingers away from a flame. These lessons were indelibly recorded in our memories precisely because they were essential to surviving the uniquely prolonged dependency and vulnerability that characterize human childhood.

FEAR AND ITS MINIONS

In *Scared Sick*, "fear" is defined as our most fundamental emotional and physical response to a perceived threat, triggering the chain of physical responses commonly known as *fight-or-flight*. Fear is recognizable in all animals, even in rodents. Rats freeze on the spot, immediately ceasing exploration and investigation. Humans are subtler, but not by much. We hold our breath, our hearts race, and both our blood pressure and muscle tone increase. We may break out in a cold sweat. In extreme cases, as when we have experienced abuse as children or have grown up in the wake of disasters such as war, ethnic cleansing or famine, our brains become permanently wired for survival in a dangerous world. Ironically, the very defenses developed to protect us under dangerous circumstances may become a huge liability in later, less-threatening chapters of our lives, especially if these defenses are now triggered without conscious awareness or control.

So, for example, the child who is constantly ridiculed or shamed or hit by an alcoholic parent and who develops extreme hypervigilance and readiness to fight at any provocation may appear aggressive, hostile, even paranoid in a less-threatening environment like school. Or a child rendered powerless in the face of adult aggression may appear frozen, depressed and abnormally passive, even self-destructive, with unfamiliar adults outside the home.

But without our adrenaline-driven alarm system, the fight-or-flight response—or more accurately, the *fight-flight-freeze* response—we would be in constant peril. When, for example, we see a snake in the grass or a swarm of yellow jackets coming for us, this is the system that enables input from the senses to instantaneously signal large muscles in the arms or legs to fight or flee, bypassing the normally slower route through the analytical brain. When there is no time for analysis, no time for deliberations concerning the perceived threat, our brain has an emergency route directly to the alarm center, the amygdala, so that in a nanosecond our entire body is activated in fight or flight. Our brain is so good at this instantaneous preparation that associated sensual information—sights, smells, sounds, recorded at the same time—can trigger fight-or-flight before we are consciously aware of the threat. But when we are helpless, like a small child in the face of a violent assault by an adult, we can neither fight nor flee. In the wake of overwhelming fear, we enter a state known as *freeze*. In the animal kingdom, freeze imitates death, allowing an animal to escape a predator. For humans, freeze works differently. We become emotionally numb, removed from reality.

The problem is that fight-flight-freeze, which originally evolved to protect us in the face of an occasional acute physical danger—like an attack from a wild beast—has not adapted to the challenges of life in the twenty-first century. Acute physical threats are no longer our primary threats. In Western culture, for example, the need to hunt for our next meal has been replaced by the need to drive on crowded freeways to stores and offices, to make money to purchase necessities, and to interface with all kinds of people, often at a merciless pace set by the technologies that now rule our lives. Having evolved as hunter-gatherers in small mobile communities close to the land, we now find ourselves living mostly in densely populated areas in constant proximity to strangers, often with little connection to the natural environment. For most of us, challenges have shifted from immediate physical threats to chronic emotional ones. To the degree that the realities of modern living—including staggering advances in technology, increasing population density and drastic changes in our roles and relationships with other humans—have outpaced the adaptations of our internal physical systems, we experience what we call *stress*. To top it all off, we are the only species, as far as we know, that worries, projecting concerns into the future and ruminating on our fears, which keeps the stress cycle running overtime.

Ephemeral sensations we call "feelings"—our emotions—fuel the stress response. In fact, our feelings, often disguised, repressed or denied, are in constant chemical communication with our brains—and consequently with all key systems in our bodies—about the status of our health and safety. When we experience a feeling of overwhelming fear, our bodies reflect critical changes in the systems designed to protect life. We are often confused about the identity of negative feelings: "What is this sensation in my gut (or my shoulders, or temples)?" "Is this anger or frustration?" "Am I anxious or depressed, exhausted or sad?" What most of us know for sure when we are "stressed" is that we experience "dis-ease." We know that we are not at ease, that we are uncomfortable; there is both clarity and irony in this term.

Most of us know the feeling of being moderately stressed, however overused and nonspecific that term may be. "Being stressed" is commonly used to denote a vague, unpleasant sense of feeling off balance emotionally or physically. Originally an architectural or engineering term to describe the pressure on structures that might cause them to break, "stress" has become a generic term that we commonly use instead of specifically describing feelings as varied as frustration, exhaustion, anxiety, distraction, fear, embarrassment and anger. When we say we are "stressed out," we might mean that we are fighting with a partner, or feeling overwhelmed by work or kids or school, or exhausted by too many demands and too little time. Regardless of vague descriptions, these negative feelings are registering in our bodies—for better or worse—chemically and organically. And extreme stress is measurable in physical terms: degree of abdominal fat, waist-hip ratio, baseline blood pressure and measures of our overnight production of cortisol or adrenaline.

In *Scared Sick*, we use the term *stressor* in reference to an external event *outside* of our bodies that results in the negative emotions and accompanying physical sensations *inside* our bodies that we call "stress." Not all stress is harmful. We experience some stress getting up in the morning, or going to school or work. For a child, receiving an immunization, going to the dentist, or getting a haircut are typical stressors with positive outcomes for the child. Some stress is essential—for example, scheduling and arriving on time for appointments, taking tests, or going for physical examinations. Researchers refer to this type of stress as *positive stress*. Positive stress actually improves immune function and facilitates an effective response to more serious stressors in much the same way that short regular sprints prepare us for the marathon. It sharpens our attention and enables us to remember life-protecting information like a mistake in judgment that we don't want to repeat. It heightens acute sensual focus. Think of driving alone at night on an icy road. It is stress—the fight-or-flight response—that heightens our alertness, sharpens our senses, and speeds our responses to the sheen on the road or the slip of a tire that indicates danger. Brief episodes of stress are what our stress systems are designed for and may actually be better for us than no stress at all.

A second category of stress is *tolerable stress*. This is stress that could become harmful, like getting divorced, having a parent or partner die, or losing a job. The capacity to recover is what keeps tolerable stress from becoming *toxic stress*—or trauma. Under tolerable stress, we have access to the healing process through relationships with friends, family or professionals and practices like regular exercise, meditation, healthful eating, adequate sleep and personal time to regroup. Though we are still affected by stress, we are able to regain internal balance or what researchers call *homeostasis*—a healthy balance within our central nervous, immune and endocrine systems that protects health. Being able to trust, to talk openly, to be heard empathically, to physically release stress through dancing or running or swimming or drumming—these are critical elements in preventing tolerable stress from turning into toxic stress.

Toxic stress is the problem. When it is strong, frequent or prolonged by emotional experiences that overwhelm homeostasis, toxic stress triggers the freeze response. In the grip of toxic stress, we don't fully regain our former equilibrium because the healing relationships and practices that may have worked with tolerable stress are now inaccessible, insufficient or unsuccessful. If it continues and accumulates in our bodies, toxic stress dysregulates the systems that protect health, paving the way for disease.

HOMELAND SECURITY: THE HPA AXIS

We all know the feeling: our blood pressure and breathing increase to mount the battle, our muscles tighten so we can run faster, leap farther or hit harder, and our senses go on red alert so we can see and hear more acutely. And as soon as the danger passes, we return to normal—collapsing or breathing a sigh of relief.

This is the normal response when one form of stress or another triggers the HPA axis: the relationship among the hypothalamus (H), the pituitary gland (P) and the adrenal glands (A) that produces finely tuned chemical messages that connect the central nervous, endocrine and immune systems. HPA is the linchpin that activates the body's main defenses. Together these three systems are the sentinels of health, functioning like internal radar. Constantly responsive to internal and external threats—from germs to terrorists—the systems of the HPA axis marshal the troops to defeat the threat, then resume their posts.

Dr. Bruce McEwen, professor of neuroendocrinology at Rockefeller University in New York and a prolific researcher on the subject, views stress as any physical or emotional challenge to the major systems of the body. He points out that the interactive nature of the three systems that comprise the HPA axis and their capacity to communicate and adjust to varying conditions have enabled humans to prevail through evolutionary challenges such as extreme climate changes, varying geographical terrains and variations in available foods. These systems work like an integrated thermostat, sending chemical messages back and forth to maintain homeostasis in response to changing conditions—especially anything that is life-threatening. Through a mutually regulatory system of lending and borrowing, activating and deactivating vital chemical messengers, they prepare the body for any perceived threat. Their job is to sustain life at all costs.

McEwen calls the HPA axis–driven process *allostasis*. This capacity to respond to a threat and return to homeostasis is essential. In healthy people, allostasis occurs almost automatically, frequently and expediently after normal physical stressors like running, chasing a ball or climbing a set of stairs. Allostasis also kicks in under everyday emotional stresses, such as giving a report in school or taking a driver's test. But problems arise when intense stressors come at us so frequently that allostasis can't fully shut down the stress response or when we need the activating energy of the stress response and allostasis doesn't shift into gear.

To imagine how allostasis works, picture a temperature gauge on the dashboard of a car. When the engine is functioning within the normal range for which it is designed, the needle stays in a green area left of center on the gauge. But if the engine overheats, the needle goes into a red area right of center, indicating alarm. Our stress response system works similarly. When we are stressed, allostasis quickly sends our HPA into the "red zone." As soon as the threat has passed, allostasis sends our alarm system right back down into the "green zone," where it is meant to function. But if the stress response is stimulated over and over without much respite, toxic stress can wear this system down so that our HPA doesn't fully return to the green zone and stay there. With chronic overstimulation, the resting state of the HPA system gradually climbs to a higher and higher default, edging toward and then staying in the red zone. Eventually it may remain there, never fully recovering its original balance.

McEwen uses the term *allostatic load* to refer to the wear and tear on the body from the overuse of allostasis and the consequent dysregulation of affected systems. When stressful conditions continue and accumulate without repair over time—or even when we neglect healthy balancing practices, like getting adequate exercise—our allostatic load will ramp up. Those of us who live in a family where there is continual conflict or fear, in a war zone, or in a violent neighborhood, or if we live in poverty or in a disaster zone—like the Japanese after the earthquake, tsunami and nuclear catastrophe of March 2011—we are at risk of chronic stress, a reality recognized by health professionals and researchers for some time.

But the surprising news is that people who experience low levels of constant annoyance every day, such as frustrating, boring or demeaning jobs, are also at risk. McEwen tells the story of workers on an assembly line at a Volvo factory in Sweden who were demoralized and miserable from doing the same job over and over. As measured by blood cortisol levels, waist-hip fat ratios and production of adrenaline, their stress levels were high. When the factory reorganized so that everyone worked on teams and on a variety of tasks, stress levels declined.[1]

McEwen believes that chronic stress is toxic stress and that it causes problems with memory, premature aging and overstimulation of nerve cells. It often leads to the loss of tissue in key parts of the brain, especially the memory center, and to dysregulation of normally protective response systems, such as the immune response. Cortisol, the hormone produced by the adrenals, calms the system and enables it to return to homeostasis, signaling the immune system that all is well. Over time *hyper*-arousal may lead to a state of *hypo*-arousal (under-arousal) as adrenal glands become exhausted and cortisol decreases. When cortisol is depleted, overactive immune responses may attack normal processes and tissues, resulting in autoimmune conditions like lupus, various allergies or chronic fatigue syndrome.

Chronic stressors are likely to be the culprits at the root of many diseases. We are especially vulnerable to chronic negative emotions (fear, anger, shame, guilt, embarrassment, grief) when we are young, particularly if we were exposed to prenatal stress. And the gloomy news is that those of us exposed early become more—not less—vulnerable to the effects of chronic stress as we age, which in turn contributes to cognitive impairment and dementia. When high levels of stress, especially worry, continue into later life, they can cause shrinkage to the hippocampus in the aging brain, reducing memory and increasing the risk of Alzheimer's. One Montreal study showed that the hippocampi of older people whose stress hormones rose over a five-year period were 14 percent smaller than in people of the same age whose stress hormones were not elevated. The former group had difficulty remembering lists of words and paragraphs and negotiating mazes—symptoms predictive of increased risk of Alzheimer's and diabetes.[2] While researchers have long known that trauma appears to contribute to the accumulation of the neurofibrillary tangles that are characteristic of Alzheimer's, they now suspect that excessive HPA stimulation in everyday stress may contribute to the disease as well.[3] So there is good reason, regardless of our age, to begin lowering stress levels; the advantages of doing so only continue to increase as we mature!

TRAUMA: FROZEN FEAR

At the farthest end of the fear spectrum is emotional *trauma*, which occurs when we are faced with either a single overwhelming event or a "last straw" in an accumulation of experiences over which we feel we have no control. Trauma is an extreme form of fear accompanied by a state of perceived helplessness—and often hopelessness. Although we tend to think of trauma as a huge event, like an automobile accident or the death of a loved one, here we use a definition similar to that of writer Annie Rogers, who defines trauma as any experience that "by its nature is in excess of what we can manage or bear."[4] Thus, when fight-or-flight, our first response to stress, fails or is unavailable, we move into the freeze response, the defining sign of trauma. In this book, "trauma" is used interchangeably with "terror" and "shock." The central difference between toxic stress and trauma is that *trauma always triggers the freeze response.* When we are unable to fight or flee, freezing is the only option left.

Trauma may not appear horrific to the onlooker. It may be subtle and quiet, and to anyone who asks, we who have been traumatized may say that we are "fine." But following trauma, our look is typically one of shock, and the state we are in is called *dissociation.* From the inside looking out, we are emotionally numb, perceiving our surroundings through a fragmented and distorted lens. Unable to respond to normal conversation with any real focus on the content, our responses may seem abstract. We may act in a rote, mechanistic manner without spontaneous affect, or suddenly erupt violently. Some victims of trauma seem amnesiac. One client whom I saw shortly after her husband's suicide (he shot himself in their home while she went for groceries) was preoccupied about the water heater that had flooded in the same room and fretful about how she was going to handle that. She was a very bright and organized thinker but appeared unfocused and repeatedly assured me she was "fine."

Immediately after experiencing trauma, most of us will find it hard to express ourselves in words; what we say may not reflect our normal range of analysis, and we may not remember what we just said. This is the impact of the internal chemistry of trauma—nature's temporary but effective anesthetic that renders us unable to feel the impact of the event, insulated from the here and now. We are in a "fog"—a state in which normal rational thinking and emotional and physical pain are temporarily suspended.

Both toxic stress and trauma trigger the same initial biological response: the HPA axis. But when trauma occurs, the energy that would normally be dispersed through fleeing or fighting is trapped. The systems in the body being primed for action are simultaneously blocked from discharging that energy, and the result is a cavalcade of maladaptations in the immune and endocrine systems. Rather than discharging the energy that accompanies fight-or-flight and returning to balance or homeostasis, two systems in the body are now at war with each other. One system activates the body with chemical signals of alarm. A second system is trying to counter this by secreting cortisol to gentle the fight-flight activation. Disease can result from the overproduction of hormones by either system. Exactly how this renders any one of us sick, emotionally or physically, depends on our genetics, prior experiences, our interpretation of the experience and the availability of support.

We are learning more about this daily. For most of human history a live look at the brain was available only through animal studies; autopsy afforded our only opportunity in humans. But new technologies in the last two decades have provided clear images of the actively functioning brain in real time. Positron-emission tomography (PET), single photon emission computed tomography (SPECT) and functional magnetic resonance imaging (MRI) allow us to see still crude but graphic changes in the brains of individuals diagnosed with post-traumatic stress disorder (PTSD). The clearest currently discerned difference is shrinkage in the hippocampus, likely resulting from excessive cortisol.

When it comes to trauma, physical and emotional are inseparable. What happens to us emotionally happens to us physically, and vice versa. And while stress and trauma are on the same continuum, they are not the same thing. Stress is a normal response to feeling threatened or overwhelmed. Trauma, on the other hand, is toxic stress frozen into place in our brains and bodies, where it reverberates chemically, generating tiny pathological shifts in immune and endocrine functions. According to a website devoted to healing trauma: "If we can communicate our distress to people who care about us and can respond adequately, we are in the realm of stress. If we become frozen in a state of active emotional intensity, we are experiencing an emotional trauma—even though we may not be consciously aware of the level of distress we are experiencing."[5]

Trauma—especially for children who have not yet learned language—tends to be stored in the brain not primarily as a conscious, rational, language-based experience in *declarative memory* but rather as a somatic or "feeling" memory stored in unconscious or *procedural memory*. Somatic memories may surface later in life in the form of physical symptoms that seem to have no discernible cause, such as chronic pain, headache or fibromyalgia.

ANXIETY: THE SHADOW OF FEAR

Humans are the only species that can develop emotional pathology based on stress because we are the only species whose advanced brains allow us to keep stressors actively present in our minds. A lingering low level of fear known as worry or *anxiety* keeps stressors alive in our minds and consequently in our bodies. Anxiety or worry—essentially the same processes—are the shadows of fear, a feeling that lingers long after the initial threat has passed or that looms long before the anticipated event. We can't stop thinking about the exam next week, or the job interview, the mistake we made, the deadline, or the possible surgery. Even when it's over, we still can't relax.

Over time, chronic anxiety may grow from being a state—an immediate but passing sensation—to being a trait, a persistent way of being. Anxiety can suspend an individual in pervasive low-grade fight-flight limbo, potentially triggering panic attacks and numerous physical conditions.

There is little question that some of us are more prone to anxiety than others. The degree of our reactivity to various stressors and the degree to which the emotion of fear can trigger or aggravate disease may be determined in part by genetics; inborn temperament seems to be a major player in the equation. But increasingly, early experience is being credited with playing an equivalent if not stronger role. Many researchers believe that "inborn" temperament is itself greatly influenced by experiences in the womb that shape who we become, anxious or otherwise.[6]

Focusing on this question of the genetic roots of anxiety, Kenneth Kendler, a psychiatric geneticist at Virginia Commonwealth University in Richmond, turned to identical twins—who share all their genes—and compared them with fraternal twins, who share only some. Kendler found that while identical twins are slightly more likely to be similar than fraternal twins in their degree of anxiety (measured as generalized anxiety disorder, panic attacks and phobias), the overall likelihood of heritability of these disorders was only in the moderate range (30 to 40 percent). He believes that our upbringing and experiences are the pivotal factors in determining our tendency toward anxiety.[7]

CHAPTER 3

Scared Sick

How Experience Becomes Biology

Experience can become biology.
—DR. BRUCE PERRY

YOU ARE ABOUT TO understand how it is that fear can kill you. We've all heard people say they are "scared to death" or "scared sick" or "scared out of their minds." We use terms like "heartsick," or "brokenhearted," or "heartache." Do these colloquialisms describe truths or are they only quaint artifacts from our primitive past? What about "dying of a broken heart"? These questions were hot topics at a conservative mainstream medical conference in Cleveland in 2007. The keynote address was titled "Voodoo Death Revisited: The Modern Lessons of Neurocardiology." Another paper was on "The Broken Heart Syndrome." Both presenters validated the medical truths behind these and other experiences that until now have typically been ignored in mainstream medical education. So how do medical researchers explain such seemingly far-fetched events? Is there such a thing as being scared to death—or voodoo death?

Dr. Vincent Felitti tells the story of a girl who went to the City Hospital in Baltimore, the hospital where he was working more than twenty-five years ago. She was born in the backwoods of the Okefenokee Swamp and delivered by an elderly midwife who made a prediction at the time of her birth. Having delivered two other babies that day, the midwife pronounced that all three of the babies were going to die early: the first would die before his eighteenth birthday, the second before his twenty-first birthday, and the third one, the girl Felitti met, before her twenty-fifth birthday.

The first boy died in an automobile accident a week before his eighteenth birthday. The second boy actually made it to his twenty-first birthday. He was so thrilled that the hex had been lifted that he threw a party for his friends at a local roadhouse. During the party, however, an armed robbery occurred; when a wild shot was fired, the birthday boy was hit in the head and killed.

Just before her twenty-fifth birthday, the girl arrived at the City Hospital in Baltimore, convinced she was doomed and seeking protection. Hysterical, she was admitted without a diagnosis, which, according to Felitti, was a major faux pas for a young medical student:

> So all we know is that we have this woman who is terribly frightened. Something is unconsciously being transmitted to various physicians so that we, who all knew better, let her into the hospital without a diagnosis. I was the second-year resident overseeing this operation at the time. . . . According to the patient in the next bed, the young woman suddenly sat up in the middle of the night and screamed, "Oh my God, it's coming!" and fell over dead. An autopsy showed inconsequential findings, nothing to support why a young woman would die. As a result, I became quite interested in voodoo death![1]

And so will you. Science is revealing amazing explanations for this and other phenomena formerly relegated to the realm of superstition. Once we understand the basic physiology of the nervous system and take a closer look at the HPA axis and the types of diseases that result from each stress and trauma, we will return to voodoo death.

RECONNECTING

Once upon a time, the connection between our experiences and our health was taken for granted. In the ancient world, it was a given that mind and body were indivisible. Hippocrates taught that physical health depended on maintaining a balanced life and that the mind was an integral part of healing disease. But in the 1500s and 1600s, following Descartes, an insistence on logic and empirical evidence as the sole measures of validity overtook the Western world; direct observation became the scientific standard. Ideas were provable or they were not. "If you can't see it, it isn't real" became a fundamental assumption of modern medicine. Emotions were split from disease, and any thinking to the contrary was dismissed as superstitious or, as with Eastern medical practices, disparaged as primitive and unscientific. Scientists shied away from exploring mind-body connections for fear of being labeled charlatans. For more than half a millennium, spurred by the invention of the microscope, Western medicine studied disease in terms of specific physical organs and chemistry, and medical treatment of each bodily function evolved into its own specialty addressing the pathological symptoms of the organs involved.

Now the tide is shifting. Despite amazing advances in technology, patients are increasingly dissatisfied with the dehumanizing effects of specialization. Many of us have deplored the lack of communication between specialists, the loss of an intimate relationship with one trusted practitioner, the endless barrage of lab tests and prescriptions for relief of symptoms that may not get to what we intuitively sense is the root of the problem. We feel reduced by managed care systems to diagnostics and a set of pathological labels, our larger selves dismissed in the process. Increasingly, we are managing our own health by incorporating Eastern practices and searching for holistic treatments.

Paralleling this push for alternatives is an explosion of new imaging technologies that enable scientists to physically observe and trace pathways shared by multiple systems, which had previously been viewed discretely. As a result, medical research is finally beginning to reveal what the ancients knew and what we ourselves have instinctively suspected: biological systems do not operate in isolation from each other.

Recognition of the integrated nature of the body's major systems has led to major shifts in Western medicine, as reflected in the term created for the newest area of study at the National Institutes of Health—*psychoendoneuroimmunology*, or "PENI," the name of both a new department and a new journal. And *neuroimmunomodulation*—the study of the connections between emotions, behavior and health—is an emerging specialty in Western medicine.

A CLOSER LOOK AT "HOMELAND SECURITY"

The central nervous system and the endocrine and immune systems communicate constantly with each other to maintain "homeland security"—or homeostasis. But at moments of perceived threat, these same systems must respond almost instantaneously. In a nanosecond they coordinate carefully titrated biochemical changes and defenses through continuous feedback loops, share resources and energy with each other, and quickly borrow and pay back debts in an ongoing dance to maintain life. It's unnecessary to remember all the terms or sequencing of this far from simple process. But pay attention to the key relationships, because it is the hijacking of these same ingeniously evolved relationships that paves the way to disease.

When you perceive a threat through one or more of your senses—say, a dog bounding toward you, growling and snarling, his teeth bared—the sight and sound and even the smell of the dog comes in through your eyes and ears and nose to an area in the base of your brain called the *locus coeruleus* (the "blue center"). The locus coeruleus immediately alerts your limbic brain—especially the *amygdala* (the "olive"). It simultaneously secretes *norepinephrine*—an energizing and activating neurotransmitter—to increase attention and vigilance in your brain and body. In less time than it takes to blink, the limbic system performs an emotional analysis and memory review of the information. So if the growling dog runs past you toward another dog, or if, as it gets closer, you recognize the dog as familiar or friendly, in milliseconds the feedback to the locus coeruleus dampens the secretion of norepinephrine and the systems readying for "red alert" in your body immediately calm.

But if the limbic review confirms the threat, just as quickly your *autonomic* (think "automatic") *nervous system* will go into action, triggered by the amygdala, which functions like the brain's smoke detector. Joseph LeDoux at New York University refers to the amygdala as the "low road" because when immediate action is required, this tiny region of the brain bypasses the "higher" or cortical brain by directly signaling alarm through the lower brain structures to the body below the neck, commanding a life-saving emergency route.[2] When there's no time for analysis—no time to linger in the grass and contemplate the color or speed of the oncoming dog—the amygdala provides the mechanism for urgent action. The "high road" through the *cortical brain*, the seat of reason and analysis, is, after all, for those times when we can sacrifice speed for analysis. The amygdala is also the center for emotionally laden memories; it communicates immediately with the *hippocampus* (the "sea horse"), the center for declarative (conscious, verbal) memory, to update this bank of emotional experiences for future reference. Since the most fundamental goal of the brain is to sustain life, these steps are crucial: first to discern the threat, then respond to it, and then record the associated variables for future protection.

Let's back up a step to look more closely at what is happening inside your brain and body to prepare you to meet this threat. Your nervous system is divided into two parts: your *central nervous system*, which consists of your brain and spinal cord, and your *peripheral nervous system*, which connects your central nervous system to your sensory organs, visceral organs, and muscles, blood vessels and glands. Your peripheral nervous system in turn consists of two systems: the *somatic system* is conscious and you have control over it, while the *autonomic system* operates outside of your awareness. It is the autonomic system that is activated by the dog—especially if the dog turns out to be a very real threat.

In such an emergency, with almost no conscious thought on your part, the autonomic nervous system, having been activated by the chemical alarm sent from the locus coeruleus (in the brainstem) and the amygdala (in the limbic brain), takes command over the battle. This system coordinates defenses throughout the body by regulating the body's visceral organs and communicating through the nerve tissue running through the heart, the smooth muscles of the body and the glands—all of which are summoned to duty during the battle to preserve life.

At this point three branches of the autonomic system start working together synergistically, like divisions of an army deployed for war. The first of the three, the *enteric division*, which runs from the esophagus to the anus, shuts down all energy that would normally be directed to the digestive process. Gastrointestinal secretion and activity is postponed until further notice, and the sphincters contract. This is no time for digestion or elimination. Reproductive and sexual activity is similarly dampened. No time for that either! The enteric system sends messages to the brain about the sensations taking place in the gut, the heart and other organs, providing what Dr. McEwen calls a set of microphones back to the brain so that it can "hear" what is going on in the body.[3]

The amygdala will have discharged a second branch as well: the *sympathetic division* initiates the fight-or-flight response, beginning with a message to the *hypothalamus* in the brain. This is the first step in activating the HPA axis. The hypothalamus sends two sets of chemical messages that act simultaneously and in counterpoint with each other: one mobilizes the main defenses of the body for fight-or-flight, while the other keeps that mobilization in balance so that the body's other tasks, such as scanning for infection, are not weakened.

The first of these messages goes from the hypothalamus (H) to an area in the center of the *adrenal glands* (A) (which sit on top of the kidneys), triggering the production of *adrenaline* to instantly stimulate the cardiovascular and nervous systems. Now the heart, lungs and large muscle groups in the arms and legs are all activated; the stress response is fully under way. Heart rate and blood pressure increase to enable the mobilization of blood to the muscles to fight or run, bronchial passages widen to increase the flow of oxygen to all points needed, blood vessels constrict to slow bleeding, glands liquidate stored carbohydrates into blood sugar for energy, and pupils dilate. Even the immune system gets involved: white blood cells that fight infection attach themselves to the walls of the blood vessels, ready for departure to any point of injury.

At the same time that the hypothalamus is signaling the adrenals, it is also sending *corticotropin-releasing hormone* (CRH) to the *pituitary gland* (P), a small endocrine gland located in the bones at the base of the skull, near the hypothalamus. Once stimulated by CRH, the pituitary releases *adrenocorticotropic hormone* (ACTH), which stimulates the adrenal cortex (the outside layer of the glands) to produce *cortisol*, a hormone that provides a counterbalance to adrenaline. Cortisol—slower to develop and longer lasting in its effects than adrenaline—serves to control the stress response and calms the immune system's inflammatory response.

While life-preserving in response to acute stress, cortisol can play both constructive and destructive roles in several diseases if over- or under-stimulated. Without constant loops of communication between the body's systems, this emergency response can easily get out of balance. Overregulated or underregulated cycles can lead to disorders of arousal, thought and feeling and a malfunctioning immune system.

The third division of the autonomic system is deployed when the senses perceive that the war is won or almost won. This division, the *parasympathetic division*, counterbalances the sympathetic functions initiated during fight-or-flight. While the maneuvers of the sympathetic system activate the body for action, the parasympathetic division now gentles those same pathways. Known as the "rest and digest" system, the parasympathetic system, responding to cortisol, slows the heart rate, dilates blood vessels, reduces blood pressure, constricts the pupils, and stimulates the digestive, reproductive and genitourinary systems after the threat has passed. Under most circumstances, the parasympathetic system comes in only briefly to balance fight-flight. But when an event is overwhelming, the parasympathetic system may continue to be stimulated, activating the muscles of the bladder and rectum and causing involuntary emptying of these organs.

In a nutshell, the amygdala sits at the hub of an exquisitely tuned and coordinated emergency response system. In an instant it can set off a bodywide alarm, triggering powerful hormones that move the body into fight-or-flight. As the bloodstream is flooded with adrenaline, norepinephrine and cortisol, the heart begins to pound, the lungs pump, and the limbs get a strong shot of glucose. The nonemergency systems in the body—digestion and immunity—are shut down so that vital energy is directed to the task at hand. The stress hormones also create a state of heightened attention in the brain, so that the urgent message from the amygdala is recorded by the hippocampus as "Don't forget!" Now, with this memory seared into the deepest recesses of your brain, you will never face a similar incident in the future without the "flashbulb" memory of this moment.

This, then, is the stress response. But what about trauma? With trauma, a marked intensity either in the severity of the threat or in the internal bodily impact of the experience combines with *helplessness*—real or perceived. If we are helpless—as an infant or very young child invariably is—or if we *think* we are helpless, neither fighting nor fleeing is an

option. In the face of extreme threat while feeling helpless, the only options are either to remain in a state of hyper-arousal or, overwhelmed by
terror, simply switch off, like an electrical appliance receiving too strong
a current.

The "off" or "freeze" switch is thrown by the *vagus nerve*, which runs
down the body from the brain stem, through the neck, to wrap around
the heart and viscera. The *dorsal vagus* is responsible for functions as varied as heart rate and gastrointestinal peristalsis, the movement of waste
through the intestines. In reptiles, the vagus nerve slows heart rate and
breathing, enabling a creature to go without oxygen so it can dive deeply
into water to escape predators or hibernate in winter. It regulates the
heart, lungs and glands and controls the volume and width of blood vessels. In mammals, the vagus nerve mediates the freeze response. When a
mammal is unable to carry out fight-or-flight because it perceives itself to
be helpless and hopeless, it collapses immobilized into a *dorsal vagal
state*—the freeze response. In humans we call this state "trauma."

Remember that this is an autonomic—or automatic—response; it is
not under our conscious control. This is the distinguishing trait of
trauma. In trauma as opposed to stress, the parasympathetic system is
called into play early, creating an insulated state that is nature's effort to
buffer pain. Normally not activated in the stress response until after the
threat has passed, trauma triggers the parasympathetic system to reduce
heart rate and blood pressure and release *endogenous opiates*, nature's own
calming and painkilling agents. A cardinal characteristic of trauma is that
both the sympathetic and the parasympathetic systems are operant at the
same time. We become both hyper-aroused and in a fog; our "thinking"
or cortical brains are hardly available to us. We often find ourselves without words and suspended in time and space—frozen in a state of simultaneous arousal and dissociation.

In the animal world, this state is called *defeat*. Think of an opossum
"playing possum," or any animal stunned when hit by a car. The animal
looks dead, a response that may save the animal's life by duping a predator. In mammals that survive the freeze, however, there is an unconscious release of the energy that was mobilized before the freeze. Legs
shake violently and claw the air, as if they were running, as they let go of
the stored nervous energy. In humans this discharge rarely occurs, due

to inhibition of the release by the cortical brain. This difference has huge implications for emotional trauma and its treatment.

Post-traumatic stress disorder (PTSD) is emotional trauma in its most extreme form. Following the return of soldiers from both world wars, when it was known as "shell shock" or "battle fatigue," and seen more recently in veterans from the wars in Vietnam, Iraq and Afghanistan, PTSD presents bizarre and often crippling symptoms. Victims vacillate between bouts of hyper-arousal—when their capacity to cope is overcome by the tiniest challenges—and times of sitting blank and still, staring into space seemingly "not there." Easily set off by loud sounds, they may startle excessively and experience panic or fits of rage. Frightening images and feelings intrude into victims' daily lives when least expected and follow them to their beds, where these images and feelings are often most horrific, interrupting their sleep and taking possession of their dreams. Plagued by nightmares that erupt into violent screams and thrashing limbs, the brain of a PTSD victim is essentially trying to mobilize the frozen energy stored in response to trauma. These are the effects of chronic overwhelming fear on a fully developed adult brain. Imagine the impact on an infant brain when fear is the architect of its chemistry and structure from the beginning.

Particularly observed in very young children, especially girls, who are less likely than boys to either fight or flee, freeze is a response to helplessness in the face of being both hyper-aroused and cornered. Parasympathetic activation enables the individual to move from terror to a state of dissociation, thus disconnecting from a horrific reality. Following even one such experience, the young child's fight-or-flight response is reactivated by reminders or thoughts of the original event, including worrying or dreaming about it. These ruminations trigger the same cavalcade of internal responses as the original experience. If it occurs often enough, it may generalize so that even subtle reminders—just fragments of the original event—are enough to trigger the full HPA response, restimulating the child's sense of helplessness each time.

When chemical states of fear persist over time in early development, the chemistry of trauma can become permanently set. Such children will always be on red alert for signs of danger. HPA systems that are constantly being overstimulated by internal or external reminders pull a child's attention away from other forms of learning. These become the children who can't sit still in school because they are busy subliminally monitoring the environment for signs of danger rather than calmly listening to the teacher. They will often perceive even benign behaviors as hostile—and they are ready to respond. Or they become the kids who don't do what's asked because they simply aren't there. Teacher talk doesn't penetrate these little brains, which have tuned out and gone away to safe places inside themselves.

Because the memory of early trauma is frozen in the limbic brain of a young child as a somatic or emotional feeling, stored without words, it will most likely not be accessible either through language or rational thought when these abilities develop. Early trauma is often at the root of physical and behavioral symptoms that defy diagnosis. Our schools are filled with children misdiagnosed with attention deficit disorder (ADD) or attention deficit/hyperactivity disorder (ADHD) whose problems stem from trauma, often related to child neglect or abuse. The symptoms look very similar—the child doesn't listen, doesn't focus, is up out of his seat, is restless, disturbs other children, or picks fights. He or she is irritable and tends to provoke or reenact violence. Twenty years ago, teachers reported that they had two or three of these children in their classroom. Now we hear that they have five or six or seven.

Negative emotions generate an ongoing stress response. Esther Sternberg, chief of the NIH section on Neuroendocrine Immunology and Behavior and author of *The Balance Within*, explains the liability of this overstimulation as the "dose effect . . . some is good, too much is bad." The dose effect in biology is likened to an inverted-U-shaped curve. Graphing the effect of stress, we find that, as hormonal levels increase with acute stress, performance improves, forming the rising arm of the inverted-U. But if acute stress becomes chronic, performance declines, forming the descending arm of the inverted-U. Sternberg asserts that this is a basic principle in biology, applicable to our bodies in many ways, including our consumption of food and drugs.[4]

So how do our emotions hijack our autonomic nervous system to facilitate our demise? Overuse of the stress response eventually undermines the very organs it is designed to protect. This happens without conscious awareness because the brain structures that handle survival evolved long before the *neocortex*, the seat of conscious awareness, and they easily override it.[5] The linchpin is the amygdala. When overstimulated, the amygdala triggers a bodywide emergency response at the expense of other vital systems, and because the episode is recorded in the hippocampus as an emergency, it isn't easily erased and forgotten—in fact quite the opposite. The hippocampus itself may be the site of extensive collateral damage if this system is activated too frequently and intensely. Let's see how this plays out.

DISTINGUISHING DISEASES OF STRESS FROM DISEASES OF TRAUMA

Beginning with Hans Selye in 1936, researchers have asserted that excessive exposure to stress contributes to the development of certain diseases. Most of the diseases that Selye identified as stress-related reflect elevated levels of adrenaline and cortisol and modifications in the HPA axis. The connection of certain diseases to stress is relatively easy to observe and evaluate. Rats exposed to prolonged and excessive stress develop digestive and cardiovascular symptoms. They gain weight, develop diabetes, ulcers, and cognitive impairment, all of which and more have been validated in human subjects over the past sixty years of research.

The consequences of trauma are subtler and have taken longer to identify and evaluate, especially in human subjects, and particularly when the traumatic experience occurred before conscious memory. Dr. Robert Scaer, who has written several books on the subject, distinguishes diseases of trauma from those of stress. He believes that diseases of trauma are uniquely characterized by symptoms that reflect the frozen pattern of cycling between parasympathetic and sympathetic responses. Scaer, who is outspoken about the hidden nature of trauma, agrees with Felitti that physicians are poorly equipped to discern trauma in pediatric patients, let alone in adults. Past trauma is masked by the metamorphosis of the event into physical symptoms and behaviors that are typically attributed to hypochondria or mental illness or character weakness on the part of the patient. Trauma is even more deeply hidden when it is rooted in earliest childhood. Whereas stress is readily observable and seen as external to the patient, trauma and its symptoms are often seen as being of the victim's own making.

DISEASES OF STRESS

We now know that negative emotions generated by stress can trigger immune responses similar to those generated by invading germs or bacteria. Emotions such as fear, anger, grief, shame and chronic frustration can stimulate an all-out immune defense with the same cascade of internal responses that the body deploys against physical pathogens, potentially dysregulating the HPA axis and facilitating disease. Keeping these systems on "red alert" can wear out heart muscles, arteries and veins, leading to many forms of heart disease and hypertension. Like overuse of a car's emergency brake, which is designed for sudden, brief threats, the overuse of fight-or-flight responses can cause immune functions to wear thin, leaving the body vulnerable to infections. Chronically overstimulated immune responses can also cause the system to attack the organs, causing autoimmune diseases like lupus and psoriasis, or they may catalyze inflammation at various sites in the body, resulting in conditions such as osteoarthritis, fibromyalgia or irritable bowel syndrome.

Too Much of a Good Thing

While *acute* stress actually bolsters the immune response, *chronic* stress does the opposite and is particularly damaging in early life, when it shapes the brain's chemical and structural patterns. To understand how this works, it is helpful to revisit the role and impact of the two key hormones involved in the stress response. Let's go back for a moment to the snarling dog coming at you with teeth bared. Remember that adrenaline is the first hormone that kicks into action. We have all heard stories of amazing physical feats when someone or something that we love is threatened—an old woman who suddenly has the strength to lift a car off her cat, or the mother who takes down a man twice her size when he threatens her child. This is the positive power of the adrenaline rush in response to danger. But if called upon too often, adrenaline causes all of the sympathetically driven processes to get out of hand.

At the first terrifying sight of the dog rushing toward you, your body's goal is to prepare you to fight or flee. If stress becomes chronic, it is the very systems engaged in mobilizing for action that are damaged. Your cardiovascular system is an obvious one. As your heart beats faster and harder, your blood pressure increases, potentially weakening and clogging your coronary arteries. Adrenaline also raises blood glucose to supply the brain and the muscles and organs for action. If the encounter with the dog is just one in a series of stressful episodes, sustained levels of increased glucose may lead to insulin resistance and set the stage for type 2 diabetes. In addition, both adrenaline and cortisol play a part in increasing lipid levels in your blood, potentially increasing cholesterol. Cardiovascular disease, stroke and diabetes are common consequences of excessive adrenaline.

While adrenaline alerted and readied your body for the threat from the vicious dog, cortisol was simultaneously circulating to provide balance so that other bodily functions—especially your immune function—would be maintained. If you subsequently realize that the dog is running past you but is headed for the throat of another dog, cortisol will simultaneously trigger the parasympathetic system to gentle the initial surge of adrenaline. Since the urgency of the danger has abated somewhat, you do not need the extreme cardiovascular and muscular responses that were initially needed. Now your body needs not only to prepare for a time of ongoing readiness for danger but also to provide the energy to sustain that mobilization. As you move away from the fighting dogs, the need to respond to an immediate threat to your life shifts to a need to get away from the scene. Or, if you know the dog being attacked, you may want to help fight off the attacking dog. In either case, shifting to a sustained effort to fight or flee requires the retention of salt and water for longer-term use and an increase in glucose to supply you with the necessary energy, whether you run or fight off the attacking dog or defend yourself if the dogs turn on you.

But wait! Now the owner is calling his attacking dog; the dog turns to return to his master, and the other dog runs off. As the crisis passes, though shaken, you go on your way. If your system is not oversensitized by chronic stress, allostasis returns your chemistry to your normal and hopefully balanced state. If this does not happen—either because you are chronically stressed and producing too much adrenaline or cortisol or because your adrenal glands are totally exhausted from too much stress—you may be harboring the roots of disease.

Just as chronic states of adrenaline production can weaken your health, chronic production of too much cortisol can cause your bone minerals to decrease, contributing to osteoporosis and some forms of arthritis. Other diseases linked to overproduction of cortisol include diabetes, functional gastrointestinal disease, Cushing's syndrome, depression, anorexia nervosa, obsessive compulsive disorder, panic disorder and hyperthyroidism. Both chronic alcoholism and excessive exercise have also been linked to too much cortisol. Cortisol activates the liver to convert fat into that burst of energy needed in a crisis; in excess, cortisol will also put a spare tire around your middle. This pattern is not reserved for those of us who are obese or overweight but can also affect very thin people. Cortisol produced during fight-or-flight mutes the immune system; prolongation of this state can lead to increased susceptibility to common viruses and bacteria because cortisol causes shrinkage of the lymph nodes (which fight off infections by collecting and destroying bacteria) and/or the thymus gland (which produces T-cells to fight infections).

Oddly enough, too *little* cortisol can also be harmful to your health. When the stress response is chronically called upon or stuck "on," the adrenal glands lose their capacity to keep up with the job; when this happens, the immune system may run wild and attack parts of the body. The consequences of cortisol depletion may appear in the absence of any discernible external threat: rashes, asthma and allergic responses are typical, as well as some autoimmune diseases like lupus, multiple sclerosis and rheumatoid arthritis. Cortisol depletion also plays a role in hypothyroidism, fibromyalgia, and chronic pain and fatigue.

Then there is the issue of inflammation. We tend to think of inflammation rather dismissively as a reaction to a splinter or a dog bite, a bee sting or a nose full of pollen. Inflammation is the immune system's way of detecting, isolating and destroying pathogens or toxic particles. An army of immune cells and antibodies rushes to the site of the pathogen, causing swelling. Since too much swelling will damage tissue and too little will fail the task, this response has to be carefully balanced. Cortisol is key to that balance. In the absence of adequate cortisol, inflammation can run rampant and generate a literally swelling tide of diseases. Unstemmed inflammation is a major underlying cause of diminished quality of life—and demise—for many.

Finally, it appears that the hippocampus—a primary memory center in the brain—is especially sensitive to cortisol. Massive amounts of cortisol may cause overactivation of receptors for cortisol in the hippocampus, causing it to shrink. This may explain the learning deficits observed in rats subjected to stress and also the forgetfulness or weak explicit memories of very anxious people. Chronic stress can have a paradoxical effect on memory. Overstimulation of the stress response can strengthen the capacity of the amygdala to form implicit or preconscious memories while simultaneously inhibiting the capacity of the hippocampus to form explicit or conscious memories. You may suddenly experience a fight-or-flight response to a voice or a smell without having the slightest idea why this is occurring or be haunted by what is commonly called "free-floating anxiety."

Elevated levels of cortisol are likely to be the linchpin in the correlations between early child abuse or neglect, adult depression or PTSD, and reduced hippocampal volume.[6] Interestingly, adult trauma victims (for example, war veterans with PTSD) have decreased hippocampal volume on the right, while survivors of child abuse have a smaller volume on the left. In a study of thirty-two women with major depressive disorder (MDD), those who had experienced severe and prolonged childhood physical or sexual abuse had an 18 percent smaller mean hippocampal volume on the left side than those who had not experienced such a childhood. A study of depressed teenagers showed similar results.[7]

DISEASES OF TRAUMA

Diseases of trauma fall at the extreme end of the continuum that begins with stress-related diseases. Traumatic diseases are those that result from chronic perceptions of helplessness and hopelessness added to toxic stress, leading to the overstimulation of the autonomic nervous system.[8] We might think that being exposed to high degrees of stress early in our lives would inoculate us against later difficulties. But the opposite is true. Chronic stress sensitizes our nervous systems, rendering us more vulnerable to trauma later. Children who experience early abuse or trauma are particularly susceptible to its diseases. They may look for years like exceptional little "stress cadets" and are often called "resilient." In fact, disease usually surfaces around middle age, though it may have been well under way earlier; disease may also not appear until relatively late in life. Chronic stress early in our lives resets our arousal system to a more sensitive level, renders us reactive to a broader range of stressors, and compromises our ability to adapt flexibly to later life stresses. As one example, Robert Scaer believes that there is a high correlation between the current rate of soldiers experiencing PTSD in Iraq and Afghanistan and trauma in their early development. He believes that early exposure to trauma often lies behind the most extreme responses to stress on the battlefield.

Inhibition of the freeze response, inherent in humans, places a person in a sustained state of entrapment, similar in its impact to the capture and caging of wild mammals. In his courageous book *The Trauma Spectrum*, Dr. Scaer asserts that, unfortunately for our species, our cages are often cultural and of our own making.[9] We humans often live under conditions that are simultaneously constructive and destructive for our health, like working under abusive conditions to pay our bills or continuing in toxic relationships to provide security for children. Such situations often generate prolonged hopelessness that has as disastrous an impact on our health as an acutely traumatic event.

Back to that dogfight you were hastening to avoid. If you have no previous history of toxic stress, that event might register in your brain as stressful but would be unlikely to kick you into trauma. You would be more likely to be able to separate the dogs and then go your way with a good story to tell—hopefully one with a happy ending. But if you have a history of accumulated trauma, particularly from early in life, and if you are sleep-deprived or coping with loss or extreme frustration or grief, an event like this could trigger trauma. This likelihood is greatly increased if you had a previous trauma involving a dog or have a panic disorder. Although the intensity and frequency of such events play a role in catalyzing trauma, it is both the meaning of the event for the individual and the backstory that has shaped it that make all the difference.

Dr. Scaer defines diseases of trauma as those that uniquely reflect the seesaw cycling of both divisions of the autonomic nervous system. Both the adrenaline-driven sympathetic division that stimulates heart rate and raises blood pressure and the cortisol-driven parasympathetic division that returns the body to "rest and digest" are involved in the strangely cycling symptoms of trauma. The unregulated oscillation of these processes is the hallmark of trauma-related diseases. Scaer says:

> The symptoms of trauma are bimodal. You have flashbacks, panic, anxiety attacks, terror. And you have dissociation, which is numbing-out and avoidance—retreat—and depression. That's been the conundrum of the whole diagnosis of PTSD: . . . you have symptoms that are both arousal-based and freeze-based. Any therapist will tell you that if you kick off a traumatic memory, the person becomes panicky and then goes suddenly into this dissociative state. . . . As I look at the so-called psychosomatic syndromes of my patients that nobody understands, they all have parasympathetic and sympathetic dominant states: fibromyalgia, chronic fatigue, irritable bowel, GERD [gastroesophageal reflux disease]. All are characterized by cyclical autonomic dysregulation.[10]

Scaer has particular sympathy for sufferers of fibromyalgia and chronic fatigue syndrome. A neurologist with thirty years of experience in trauma and rehabilitative medicine, Scaer says these patients have been poorly understood, poorly treated, and often dismissed as hypochondriacs. It is common, he says, for practitioners of Western medicine—with the exception of rheumatologists, who specialize in these syndromes—to view these patients as suffering "psychosomatic" illnesses. The symptoms—hard to measure objectively and hard to remediate—include diffuse skeletal pain, points of tenderness across the body's surface, morning stiffness, daytime fatigue and interrupted sleep. Victims of fibromyalgia experience many fluctuating symptoms of autonomic dysregulation, including numbness, tingling, hypervigilance, emotional instability, dizziness and cognitive impairment. Scaer believes that fibromyalgia reflects preverbal trauma that is difficult to document. The muscle pain, he notes, is probably the result of tightening muscles during REM sleep in which the patient is reenacting the traumatic event. Scaer believes that the irritability, fatigue, and emotional and cognitive impairment experienced by fibromyalgia sufferers are due to the subsequent loss of restorative sleep from the interruption of the REM cycle.

The seesaw of competing functions is also evident in GERD (gastroesophageal reflux disease), a condition in which stomach acid backs up into the esophagus through a faulty sphincter. The result is pressure and discomfort from heartburn. Scaer hypothesizes that the impaired coordination of the stomach and esophageal sphincters characteristic of GERD may be caused by disruptions of homeostasis that involve the oscillation of both sympathetic and parasympathetic symptoms.

Irritable bowel syndrome (IBS) is characterized by the cycling of the gut between diarrhea and constipation. As is true of fibromyalgia, symptoms include fatigue, sleep disturbance, soft tissue pain and bladder symptoms. Scaer views IBS as the result of an ongoing freeze response that has taken hold in a vulnerable bodily system.

MVP (mitral valve prolapse), like GERD, involves the backward leakage of a fluid—in this case, blood—through an incomplete closure of a valve. In cases of MVP where there is no anatomical abnormality, disrupted homeostasis from trauma is the likeliest culprit. MVP can be accompanied by panic attacks, palpitations, dizziness, cognitive symptoms, exaggerated startle responses, numbness, tingling and chemical sensitivities—all of which, according to Scaer, reflect once again the cycling of sympathetic and parasympathetic processes. Many MVP sufferers also have IBS.

Finally, Scaer views multiple chemical sensitivities—being overwhelmed by smells or sounds or tastes or textures that would not affect most people—as a condition that echoes aspects of fibromyalgia and chronic fatigue. None of the three is associated with any objective abnormality that can be measured. Although Scaer admits that many of his conclusions are conjectural, he feels certain that time will validate his belief that each of these conditions arises from trauma—typically in early childhood and often before there is conscious memory of the event. He is a tireless advocate of prevention of child abuse and neglect, which he believes is at the root of overlooked and dismissed trauma that emerges decades later when it appears as physical or emotional disease.

Scaer uses the neurobiological concept of "kindling" to explain how homeostasis is derailed by strong exposures to trauma. In common usage, kindling is the wood that easily ignites larger pieces of wood. In Scaer's paradigm, kindling is a process in the brain whereby pathways that have been sensitized by stress can spontaneously ignite without further stimulation. Sensitized neurons take on lives of their own, spontaneously stimulating neighboring neurons with no outside event triggering the activity.[11] Neurons in the amygdala are particularly prone to this kind of activity.

The kindling concept helps explain why so many diseases associated with trauma actually get worse over time, with little outside stimulation. Internal cues, especially from procedural memory, may lie behind the "recovered memory" phenomenon—in which information from events that occurred months or years earlier surfaces over time. These symptoms can actually increase as sensitivities build on themselves. Victims of complex trauma pose particular challenges for doctors given the lack of "objective" evidence and their inability to provide rational, language-based contextual information or history. That they are often dismissed as neurotic only adds to their frustration and exhaustion.

The kindling process in the traumatically sensitized brain explains some of the most difficult diagnoses known to modern medicine. Scaer believes that this group of very real diseases is this century's "hysteria": like those suffering from the "hysteria" diagnosed in Freud's time, patients presenting symptoms of these diseases are often met with minimizing, even hostile, responses from physicians and from the larger community. Scaer has taken a controversial—and probably accurate—position on these diseases by attributing their etiology to childhood trauma, which is far more rampant in our culture than is commonly recognized. In Scaer's words: "These are all diseases of dysregulation, diseases that are chronic, insidious, and don't have an endpoint that we can measure consistently with lab tests and imaging. So we don't recognize these diseases as significant or real."[12]

What does all of this have to do with Dr. Felitti's patient from the Okefenokee Swamp? How do Scaer's hypotheses apply to voodoo death? You have probably already figured out that voodoo death has something to do with trauma. But how does it happen?

The key to understanding voodoo death—and trauma—is the dorsal vagus nerve. It is the vagus nerve that allows some animals to hibernate—like frogs sleeping through the winter in the mud at the bottom of a creek or a pond, or bears sleeping for months without food and with a pulse rate of maybe two beats a minute. The vagus nerve lowers the oxygen demands of the cells while also slowing down all of the viscera—the heart, lungs and intestines, the glands of the intestines, and the state of constriction of the blood vessels. In humans the chief function of the dorsal vagus nerve is to throw the freeze switch and regulate the freeze response: as heart rate and breathing slow almost to a stop, we may collapse or become immobile. This is the mechanism that puts us in a state of dissociation, which is meant to provide protection when all else fails. But when this state continues, especially in humans, we can sink so deeply into a state of low oxygen and slowed pulse that we die. Believing ourselves to be helpless and hopeless in the face of a curse, this extreme parasympathetic response can usher us into death . . . as in voodoo death.

Voodoo death is actually a profound freeze response: the person dies with the heart relaxed and full of blood. Though few of us in Western cultures will die a voodoo death, we are all in fact emotional creatures. We ignore this powerful understanding at our peril.

SCARED STUPID

Historically, most of us have viewed memory loss the same way we view many forms of disease, as a normal aspect of aging or an unfortunate result of genetics. Certainly genetics plays a role in diseases like Alzheimer's, but it appears that, as in the diseases discussed earlier, stress and trauma are to memory what the iceberg was to the *Titanic*. Fear plays a role in all forms of memory loss and has strong implications for learning in general. Memory—reconstructing our past experiences—requires the coordination of several brain circuits and systems. Our memories are stored categorically; when, where, who, what it looked like, feelings about the experience, are each filed away in separate areas of our brains. Recall that there are two basic types of memories: conscious ("explicit" or "declarative") and unconscious ("implicit" or "procedural"). Conscious memory is mediated by the hippocampus, which contains memories of experiences that we can access and share. Unconscious memory resides in several systems, each of which operates without our awareness. For example, learning to ride a bike or to play the piano is stored in the motor areas of the brain as procedural memory, while strong emotional experiences, particularly those involving fear, are mediated by the amygdala and processed through the orbitofrontal cortex. In adults it is normal for both conscious memory, mediated by the hippocampus, and unconscious memory, mediated by the amygdala, to work together.

But in the beginning, this is not the case; only the amygdala is well developed at birth. The hippocampus does not come fully online for another three or four years. Babies do not record conscious memories.[13] Early emotional experiences are recorded as feelings and sensual fragments: sounds, smells, tastes and ephemeral body sensations. Long before conscious memory records the contextual details of experiences, unconscious memory records emotional sensations. Then, as we mature, these two systems can work either together or independently.

Let's go back to the snarling dog one last time to see how this works. If the dog had run straight for you and buried his teeth in your leg, your HPA axis would have been fully engaged—both conscious and unconscious memory systems would have recorded the scene as you experienced it. For some time after the incident, depending on your experience with dogs before, any encounter with the sound of a barking or snarling dog or other details recorded at the time—like the sound of a sprinkler system or the color of the leather in the dog's collar—might be enough to trigger the full cascade of stress responses. At the neighborhood picnic a year later, for instance, the sound of the neighbor's barking but totally adoring Labrador retriever might trigger a response. But because you were an adult when the bite happened, it is likely that your conscious memory will kick in, enabling you to remember the context and to recognize the difference between your neighbor's Lab and the attacking dog a year earlier. You know at a visceral level that you are safe. Your body will calm, and you will return to the picnic without much, if any, obvious discomfort.

However, if you were a toddler when the snarling dog bit you, this would play out very differently. The memory would have been stored by your amygdala in your unconscious memory, and the scarier the experience the stronger the memory. As an adult at that picnic years later, hearing the neighbor's sweet Lab barking is likely to flood you with the same overwhelming fear that was stored in your unconscious memory at the time of the original bite. The younger you were when the original event occurred, the more overwhelming the barking dog will be to you as an adult.

As we mature and our experiences stimulate cortical growth, our rational skills gradually modulate the reactivity of the amygdala. Children, especially those developing in nurturing environments, become more sophisticated in screening their earlier alarms and sorting out small discomforts or harmless events from more threatening dangers. So we see the toddler laugh when a parent says, "Boo!" but cry when the same sound comes from a stranger. Parents are pivotal in enabling a baby to begin to self-regulate fear and other negative emotions. But this process of learning to self-regulate takes years and requires regular and ongoing boosters throughout adolescence.

Even if you are by now an older adolescent or adult, if the biting dog incident turns out to be just one in a series of stressful or traumatic events in your life—say, you are going through a difficult divorce or enduring a miserable job or loss of employment—the dog bite may be the last straw, resulting in some impairment to your conscious memory. The impact of stress hormones on the hippocampus from chronic negative emotion is well documented. Paradoxically, when stress prevails, the reactivity of the amygdala is *strengthened,* so that *unconscious* memories are vividly stored while simultaneously the hippocampus is inhibited in its ability to form conscious memories. This enhanced reactivity is often the root of free-floating anxiety that can lead you to be frightened by sensual stimulation, such as being in a crowd or hearing an explosive sound—with no conscious awareness of what triggered your alarm.

The pioneering work of Dr. J. D. Bremner, director of the Emory Clinical Neuroscience Research Unit on Vietnam War veterans, provides an interesting adjunct to the study of the impact of extreme stress on memory. Bremner has focused on the role of the orbitofrontal cortex in PTSD. The orbitofrontal cortex is the part of the brain that sits just between and slightly above the eyebrows. It analyzes the fear response from the lower brain, enabling you to distinguish whether or not your feelings and awareness in a given moment are happening now, are being viewed on a movie screen, or are a hallucination.[14] A cardinal symptom of PTSD, which is characterized by alterations in the functioning in the orbitofrontal cortex, is the inability to distinguish flashbacks from real occurrences: chronically intrusive memories of a trauma are experienced as if occurring in the present moment. For this reason, PTSD is described by some researchers as a "corruption of memory," wherein each discrete sensual aspect of a previously overwhelming experience—sounds, sights, smells, tastes, sensations of all sorts—remains stored in unconscious procedural memory. Restimulated experiences continue to resonate, and the

PTSD sufferer has no ability to distinguish the present moment from the memory; all internal systems are as engaged as they were when the incident first occurred.

It is interesting to note that the orbitofrontal cortex not only undergoes critical development in early infancy but is also the part of the brain, according to Dr. Allan Schore (a leading researcher in the field of neuropsychology at UCLA's David Geffen School of Medicine), that is critically shaped by the quality of attachment between infants and their primary caregivers. Schore points out that our first relationships play a strong role in the competent functioning of the orbitofrontal cortex, which, in turn, becomes a critical part of our memory as well as our overall health and competence.

Under normal circumstances, conscious and unconscious memory systems become integrated as we develop. All of us have frightening memories: an unconsciously recorded sound or an odor may trigger the recall of a fear-saturated event, just as the bark from the friendly Labrador retriever at the neighborhood picnic triggers the memory of the biting dog from the past. Recognizing this, we can put the unconsciously recorded fear in a new context, creating for our autobiographical file of memories a new integrated conscious memory that includes formerly separated feelings. This process, which is key to moving short-term memories into long-term memory, is mediated by the hippocampus and central to trauma therapy.

Trauma permeates our lives, but the form it takes can be abstract. The memory of trauma may be stored as sensual fragments—images, sounds, smells, tastes, bodily sensations—that tend not to change over time and may emerge in daily life in strange and unrecognizable ways, including odd phobias and fears. To the untrained eye, symptoms of trauma may be mysterious and misleading. In young children, for example, trauma may manifest in play that is angry and violent. It can be hard to distinguish traumatized children from children who have ADD or ADHD or who are bipolar—to the point that some researchers speculate that these conditions are rooted in undiagnosed prenatal or neonatal trauma. Night terrors or nightmares may disrupt sleep. Children attempting to resolve frozen fear may erupt aggressively, hurting toys or themselves or other children. These efforts to externalize frozen feelings may be misdiagnosed and medicated in ways that repress the behavior but do little to heal the underlying issue, which may ultimately manifest as a physical symptom like stomachache, headache, or bowel or allergy problems.

Preconscious memories may emerge as diseases that are hard to trace to their origins and are consequently hard to diagnose. Felitti and Scaer both point out that fear is one reason why physicians may avoid exploring patient histories that are hard to talk about, hard to validate, and hard to measure. The physician's fear of being in unfamiliar territory may combine with the patient's own fear so that pertinent facts underlying the symptom are obfuscated. Nonetheless, this is the new frontier. New research into Alzheimer's and new findings in epigenetics, together with the findings of the ACE Study, all point strongly to the role of chronic stress and early trauma in memory loss as adults age.[15] It's hard not to be excited by imagining what we might gain if we began to act on the implications of this research.

Several authors have explored the relationship between fear and irrational behavior by examining historical and current governments and their use of fear to control a population. None have done this with more insight than Naomi Klein, whose book *The Shock Doctrine* examines the use of natural and governmentally created disasters to engender fear in a society purposely rendering people more receptive to aggressive solutions that they would not normally endorse.[16] One does not have to look far to see this phenomenon at work. Whether we are talking about Nazi Germany or Romania in the late 1990s or post-9/11 America, we are all subject to strong and expedient solutions when we fear for our physical well-being. Complex analysis, critical thinking and mindful judgment can be bypassed within all of our brains when we are afraid. Fear sells not only newspapers and television shows and all kinds of things we don't really need but also, most sinister of all, wrong-minded, often greed- and control-driven solutions to complex problems. Thus, the trauma cycle perpetuates itself, at both an individual and a societal level. And it begins with babies. As Dr. Perry says, "Experience becomes biology." And biology shapes history.

CHAPTER 4

Little Traumas

Prenatal and Perinatal

GIVEN THE NUMBER of diseases linked to trauma, how is it that trauma continues to have such a large reach? The physiology is clear, yet we continue to see growing rates of trauma-related illness, and it increasingly affects our young. The answer to this question lies in part in our ignorance about what trauma looks like, especially in humans too young to fight or flee or even scream. In contrast to the common interpretation of the term, emotional trauma, even in adults, is not always the result of a sudden disaster, like war or a fatal accident. Dr. Robert Scaer believes that emotional trauma, especially for a baby or young child, can result not only from a single searing experience but also from the accumulation of myriad "little traumas" that have the same effect, setting up a lifelong course of systemic dysregulation.[1] The brain's most essential role is to ensure survival. So if, through sensual perceptions or chemical messages, the fetal or newborn brain chronically detects that the world is aggressive and hostile or that survival depends on vigilance, his or her very plastic brain will shape itself accordingly, sending chemical messages to the endocrine and immune systems to prepare to survive such a world.

BABY PAIN

Only twenty-five short years ago, most physicians in the United States and across the Western world believed that premature and newborn babies were incapable of feeling physical pain. Hospital practices reflected this belief. In *Ghosts from the Nursery* we referenced an article by Dr. David Chamberlain, past president of the Association for Pre- and Perinatal Psychology and Health, in which he described a routine procedure necessary for 50 percent of infants weighing less than 1,500 grams that was commonly performed without anesthesia. The operation involved cutting a hole in the chest and in both sides of the neck, making an incision from the breastbone to the backbone, prying the ribs apart, retracting the left lung, and tying off an artery near the heart. This surgery took an average of an hour and a half, during which the baby was flooded with pain and terror. Many died from pain and shock. Yet until 1986, anesthesia was routinely withheld.[2] It wasn't until 1987 that the American Academy of Pediatrics called for an end to the practice of operating on newborns without anesthesia, or, for that matter, without any form of pain relief.

A shocking article in the *New York Times Sunday Magazine* entitled "The First Ache" featured the experiences of Dr. Kanwaljeet "Sunny" Anand, a medical resident in a neonatal intensive care unit (NICU) at the John Radcliffe Hospital in Oxford, England. In 1983 Anand became puzzled by changes he observed in the preterm babies under his care who were wheeled out of the NICU for surgeries.[3] When the babies were returned, he noticed that they were terribly stressed: their skin was gray, their pulses were weak, and their breathing was shallow. Wondering what was happening, and breaking with hospital protocol (as a pediatric intern, he was not invited into surgery), Anand followed several of his tiny patients into the operating room. What he discovered was horrifying— and was also routine practice. Babies were being cut open for major surgeries to repair malfunctioning hearts and lungs and kidneys without anesthesia! They were given only a paralytic drug to keep them still while they underwent the knife. The operant belief in Western medicine was that the nervous systems of newborns were too immature to feel pain and that the risks of anesthesia far outweighed the benefits. This amazing assumption was based on a mid-twentieth-century observation by researchers who administered pinpricks over the bodies of newborns and found that there was a barely detectable but graduated response to pain from birth to ten days of age. The researchers of the era concluded that newborns had limited if any pain perception owing to immature development of the cerebral cortex. Totally overlooked was the possibility that maternal anesthesia, routinely used for deliveries at that time, accounted for the diminished responsiveness of infants in the first few days following birth.

Observing the serious setbacks to the babies, Dr. Anand decided to test the pain assumption. In a series of clinical trials, he demonstrated that operations performed on newborns without anesthesia produced a "massive stress response," as evidenced by a measurable flood of stress hormones. Babies who received anesthesia had lower stress hormone levels, better stabilized breathing and blood sugar levels, and fewer postoperative complications. Most importantly, after hospitals started using anesthesia on infants undergoing heart repair surgery, newborn mortality rates dropped from around 25 percent to less than 10 percent. Today adequate pain relief, even for preterm babies, is the standard of care in American hospitals.

Fortunately, Anand didn't stop there. As life-saving technology has improved, survival rates for preterm infants have grown to include younger and younger fetuses. Observing babies in the NICU as young as twenty-two weeks gestational age, Anand noticed that most of them grimaced when poked with a needle. Now a professor at the University of Arkansas for Medical Sciences and a pediatrician at the Arkansas Children's Hospital in Little Rock, Anand asserts that by twenty weeks of gestation the nervous system is developed enough for fetuses to experience pain. His detractors insist that the nervous system has not yet matured to the point of enabling these tiny beings to experience real pain and that the fetal grimaces observed by Anand are simply reflexive. The debate continues as surgeries to repair defects in fetuses become more common.

At least half a dozen fetal treatment centers around the nation routinely perform fetal surgeries in order to clear blocked urinary tracts, remove lung tumors, and repair malformed organs, including those resulting from prenatally diagnosed spina bifida. Whether or not to anesthetize is a decision left to the surgeons who officiate in these centers. While the issue remains cloudy, several other physicians across the globe have echoed Anand's observations. Dr. Nicholas Fisk, a fetal medicine specialist and director of the Queensland Center for Clinical Restoration in Australia, looks back regretfully at his own early history of assuming that fetuses experience no pain. Fisk says that his response to mothers who asked him before he operated, "Does my baby feel pain?" was once: "No, of course not." His experience caused him to change his mind. Fisk's research led him to discover that fetuses as young as eighteen weeks in fact react to stress with measurable hormonal changes and that a protective flow of blood is directed toward the brain—a physiological sign of an effort to protect a vital organ from threat. These changes are identical to those seen in full-term infants and adults in response to pain. In a study similar to Anand's, Fisk observed forty-five fetuses requiring painful blood transfusions. After giving one-third of them an injection of fentanyl, a painkiller, he found that the production of stress hormones was halved in these fetuses and that their blood flow pattern remained normal.[4]

The issue of pain in fetal development is crucial to understanding the influence of trauma on human development. Denial still continues. For example, a noted psychologist at the University of Birmingham in Britain, asserts that babies can't feel pain until they are at least a year old, based on his theory that the experience of pain has to be learned from an adult. By this reasoning, a fetus, having had no contact with an adult yet, has had no opportunity to learn pain. Mark Rosen, the anesthesiologist on the very first open-heart fetal operation, performed in 1981 at the University of California at San Francisco, has set the standard for anesthesia protocols for fetal surgery across the world. He doesn't believe that fetal pain exists until the third trimester of pregnancy, at twenty-nine or thirty weeks, when the nervous system has matured enough to get pain signals up through the thalamus to the cerebral cortex. The *New York Times* article cited Rosen's assertion in a previous *JAMA* publication that pain cannot exist until the cerebral cortex, as the organ of consciousness, is fully functioning.

This theory is hotly debated. For example, Swedish neuroscientist Bjorn Merker has studied hydrocephalic children, who have a brain stem but no cerebral cortex. His findings indicate that these children still appear to have a form of consciousness. Merker believes that the brain stem can support a level of awareness without the benefit of the cortical brain. These are questions that will only be resolved as we learn more about the brain. But meanwhile, if we don't implement the appropriate safeguards to protect babies from pain, we may be increasing the rate of trauma in children through the procedures we are using to save their lives.

Anna Taddio, a pain specialist at the Hospital for Sick Children in Toronto, observed more than a decade ago that the baby boys she treated in the NICU were more sensitive to pain than the baby girls. She considered several potential explanations: sex hormones, anatomical differences, or a painful event experienced only by boys, such as circumcision. Taddio designed a study involving eighty-seven baby boys. She found that baby boys who were circumcised soon after birth cried longer and louder than uncircumcised babies when they were vaccinated four to six months later. Baby boys who had received an analgesic cream at the time of the circumcision cried less upon getting the shot than those who had been given no pain relief. Taddio concluded that a single painful event can affect babies for months, maybe longer. Like Anand, she believes that the younger the child the *more vulnerable* the nervous system is to pain, and that the baby's physical mechanisms for inhibiting pain and making it more tolerable don't become active until well *after* birth. Immature physiology may in fact render the fetus *more* sensitive to pain, not less.[5]

The debate about when real pain is first experienced remains a running controversy to this day. At risk are those children whose early exposure to pain may alter their pain threshold, making them either hypersensitive or frighteningly inured to pain and setting the stage for later psychological and behavioral problems, including anxiety and depression. Without using the word *trauma*, Taddio is clearly talking about the same brain-based shift when she says:

> When we do something to a baby that is not an expected part of its normal development, especially at a very early stage, we may actually change the way the nervous system is wired. This is precisely what many leading researchers now believe sets the stage for later disease, both physical and emotional. We are inadvertently altering the nervous systems of young human beings—almost always unknowingly and for good, often life-saving reasons—but there are consequences in later development which are typically disconnected from their roots in earliest experience and are often misdiagnosed.[6]

FROM THE VERY BEGINNING

Pain is far from the only prenatally rooted potential threat to adult health. While it is almost impossible to imagine, there is growing belief that *even before* the fertilized egg becomes an embryo, critical decisions have been made that will forever influence adult physical health. Long before it lies suspended in the liquid chamber of the mother's womb, the pre-fetus is engaged in a dynamic chemical conversation with the mother. The goal of this conversation from the very beginning is to communicate what adaptations will be needed to enable the fetus to survive any conditions that the mother's body is signaling might threaten its survival. The mother's womb previews the world for the baby, whose survival depends on this chemical dialogue. This begins so early that mothers are not only unaware of the conversation, they are generally even unaware of the pregnancy.

It is hard to believe—let alone plan for—how early this dialogue begins. Dr. David Barker and his team at Southampton University in England have discovered that even before the fertilized egg is implanted in the uterus, the human organism is taking its cues from the environment in the mother's body. Depending on the nutritional environment on the way to the womb, developmental decisions are set in place that render that individual more or less susceptible to heart disease and hypertension.[7] In just six days after conception, *prior to implantation of the fertilized egg in the womb*, crucial messages have been acquired by the zygote, then the blastocyst, that direct the course of that future individual's heart health. At this remarkably early point, the critical factor is maternal nutrition. If the environment within the mother's reproductive system reflects a lack of nutrients, the fertilized egg and subsequently the fetus will slow down its rate of growth to help it survive a low-nutrient environment postnatally. So the full-term baby will be small (low birth weight being defined as five pounds, eight ounces, or less). These responses probably improve the competitive fitness of the offspring to live long enough to procreate. But for individuals who instead happened to arrive at birth in a nutrient-rich environment, the mismatch between the environment for which they were prepared in the womb and the world they were born into will take its toll by midlife in the form of adult disease.

Barker and his associates, including Dr. Thomas Fleming and Dr. Adam Watkins of the University of Southampton, have discovered through animal studies that the physiology of prenatal undernutrition takes its toll on key organs. Baby may look normal at birth and throughout childhood. Problems do not typically surface until later in development, so that the individual has enough organ health to survive, procreate and preserve the species. But because the individual's heart, kidney, pancreas, lungs or gut may have had to sacrifice cell quantity to complete themselves without adequate prenatal nutrition and still be ready for birth, one or more of these systems may have a sort of "planned obsolescence." Like siding bought at a bargain price, organs that were small and compromised by the quality of raw materials will give out more quickly than those of a person who had adequate nutrition in the womb. Low birth weight, which generally means small organs, places humans—especially men—at relatively higher risk of coronary heart disease, stroke, high blood pressure, diabetes and obesity. Clearly, other factors will come into play, such as smoking, diet, exercise and stress, along with genetics. But a basic tendency can be cast—even before a pregnancy is recognized.

An epidemiologist, Dr. Barker discovered the correlation between gestational undernutrition and adult heart health in the 1980s when he began compiling maps that showed common causes of death in regions of England and Wales. Finding a pattern of both low birth weights and deaths from heart disease in adults from economically poorer regions, he began correlating birth and death records. Men whose birth weights were below five and a half pounds had the highest rate of death from heart disease: they were 50 percent more likely to die of heart disease than men who had begun life with a birth weight of nine and a half pounds. Through a retrospective study of Finnish adults with coronary heart disease, Barker also found that babies who start life at a low birth weight, then remain small for their age at one year but rapidly gain weight during their second year, are particularly prone to later coronary heart disease, hypertension and type 2 diabetes.[8] Barker speculates that

> rapid weight gain may lead to a high level of body fat in relation to muscle. This may explain why this growth pattern is related to insulin resistance and thus, coronary heart disease. These findings are further evidence

that prevention of coronary heart disease depends on three things. First, mothers should have a balanced and varied diet before conception and during pregnancy. Second, infant growth after birth should be protected. Third, rapid weight gain after two years of age should be avoided in children who were small at birth or thin at two years.[9]

As the basis for new research under way both in Britain and at Oregon's Health Science University, Barker's fetal origins hypothesis (widely known as "the Barker hypothesis") proposes that coronary heart disease and several related diseases are the result of fetal adaptations that were meant to ensure survival when there was compromised nutrition in the womb. Although intrauterine growth restriction is adaptive for short-term survival, adult disease is the long-term cost. Scientists are studying the mechanisms involved, but the operant belief is that maternal undernutrition short-circuits organ development via the placenta. The primary player in this internal drama is probably cortisol—the same hormone that affects the HPA axis of a fetus whose mother is severely anxious or depressed. Cortisol speeds up organ development to prepare the baby to survive when normal gestation is threatened, but not without a price.

So a mother's nutritional status shapes the placenta, and the placenta shapes the baby's organs that must be ready by birth. Take a baby's heart, for example. Prenatal undernutrition leads to fewer heart muscle cells, and therefore the individual has fewer in reserve when his or her heart is challenged in later life. Low-birth-weight babies also tend to have smaller girth around their bellies, a factor predictive of higher cholesterol later in life; Barker postulates that this reflects poor liver growth resulting from the mother's undernutrition. Lack of nutrition also affects the baby's kidneys, leading to fewer nephrons to extract wastes from the blood. All of us lose nephrons as we age, but low-birth-weight individuals, who have fewer to start with, are more likely to exhibit high blood pressure at middle age, when they no longer have enough nephrons to adequately cleanse the blood. This is especially true for an overweight individual, since a larger body has more blood to cleanse.

Prenatal undernutrition also affects other organs and functions, such as arteries, veins, the pancreas and bones. With poorly developed blood vessels in their brains, low-birth-weight babies may become more susceptible later to the normal wear-and-tear of aging on the circulatory system. Their vessels may more easily burst or become blocked at middle age, and they are more likely to have a stroke. In response to the undernourished fetus's thrifty handling of scarce sugar, the pancreas can become hardwired to flood the blood with sugar later, when food becomes plentiful; the resulting obesity contributes to insulin resistance, from which diabetes can develop. Bone mass is also compromised in low-birth-weight babies, leading to osteoporosis, and adding insult to injury is the strong correlation between osteoporosis and depression. Thus, both physical and emotional health may be at risk at midlife for low-birth-weight individuals.[10]

While maternal undernutrition may not constitute trauma in the sense of conveying overwhelming fear to the unborn, it may set the developing nervous system on a trajectory that will be particularly vulnerable to fear and other negative emotions after birth. Additionally, lack of nutrition in mothers is often accompanied by another powerful force in shaping trauma: prenatal maternal stress. Inadvertently shared within the very first relationship, a mother's emotional stress during pregnancy by itself may take a greater toll on her baby than on her own body.

WHEN STRESS BEGETS STRESS

Ancient societies often recognized the impact of fear and stress on the unborn. In primitive tribes ranging from Africa to Native America, women were warned during pregnancy not to look on unpleasant sights and to avoid frightening emotional experiences, which were commonly believed to produce physical and spiritual defects in children. In recent history, especially in Western culture, all such admonitions about the influence of the mother's emotional state on the baby have been dismissed as folklore, and indeed, some of this advice clearly falls in the category of "old wives' tales"—like the notion that a pregnant woman who is suddenly frightened by a jumping rabbit will give birth to a child with a harelip. But as women struggle with rising rates of divorce, domestic violence, single parenthood and poverty, the emerging science is forcing us to acknowledge what ancient wisdom has taught, and what many of us have intuitively known: maternal stress affects the fetus. Maternal stress during pregnancy is highly correlated with spontaneous abortion, pre-eclampsia, preterm birth, low birth weight and the adult diseases that follow these conditions. Particularly worrisome, as the world changes exponentially and women are stressed in unprecedented ways, is the role played by maternal stress in a host of later emotional and behavioral problems in children. Gestation is the time when our nervous systems are under construction and being wired for equanimity and stability or for hypersensitivity and vulnerability to the stressors of the world outside the womb—in other words, the time when we are primed for either vulnerability or relative durability in response to later trauma.

We often refer to variables such as reactivity to frustration, sensitivity to new people and situations, and emotional lability as characteristics of *temperament* and ascribe them to genetics. Temperament is influenced by genetics, but it is also in part a measure of central nervous system reactivity, which is greatly influenced by the prenate's environment. The esteemed obstetrician Dr. Peter Nathanielsz writes in *Life in the Womb: The Origin of Health and Disease* about his own observations of the variability in people's nervous systems.[11] He notes that most of us have systems that have been taken off course a bit by the trauma inherent in human experience: some of us tend to react, to lash out, allowing our more primitive limbic brain to shape our behavior. Ours is an emotionally unmodulated culture and world, and a quick look at a television news program—chronicling war, road rage, violent crime, widespread addiction—reveals our lability as a species. Increasing levels of emotional reactivity are not an asset in a world that is rapidly growing smaller and competing for limited resources and where weapons of mass destruction await instantaneous activation.

To read recent research on the influences of maternal stress and malnutrition on the fetus and the lasting impact of those experiences is like reading the scientific counterpart of an astrological chart. Our first experiences, sometimes in interaction with genetics but often with little genetic input, can create lifelong differences in our temperaments and in our health. It appears that Shakespeare was well ahead of his time, and science, when he wrote in *Julius Caesar*, "The fault, dear Brutus, is not in our stars, but in ourselves."

To establish linkages between factors like maternal stress and negative outcomes to offspring, researchers employ three types of studies. One is *epidemiological*, like Barker's early work that looked at correlations between newborn measurements recorded in birth records and causes of demise in the death records of the same individuals. A second type of study, *clinical research*, examines connections between disease and other factors based on histories, examinations and medical tests. Vincent Felitti and Robert Anda's ACE Study was a clinical work that correlated childhood histories reported through a questionnaire with medical records of disease at midlife. Finally, *animal research* affords the opportunity to control discrete processes that cannot be controlled in a human clinical or epidemiological study.

Researchers commonly use all three types of research when investigating a given hypothesis. When it comes to studies of the influence of stress on development, the ethical prohibition on deliberately stressing human subjects has led most researchers to look closely at the impact of hypothesized variables on animals, chiefly rats, monkeys, and—in the case of Barker's work on embryo development—sheep.

LESSONS ON MATERNAL STRESS FROM ANIMAL BABIES AND THEIR MOTHERS

Studies of maternal stress in rats and monkeys have been conducted for decades. Pregnant rats have been tormented with stressors varying from loud horn blasts to foot shocks to immobilization during the early, middle and late gestation of their pups. Pregnant monkeys have also endured alarming sounds and lights, exposure to simulated predators, and sudden changes in social groupings, including crowding and isolation. All of this has been done so that their offspring could be evaluated for differences in their emotional, cognitive or physical health.

In both species, comparisons of the offspring of nonstressed mothers and stressed mothers have shown clear differences in the reactivity of their nervous systems. Rat pups and baby monkeys of mothers chronically stressed during pregnancy, particularly during the early days of the pregnancy (equivalent to the first trimester in humans), have overreactive, hypervigilant and pervasively anxious responses to normal stressors after birth. More often than not, the effects continue throughout development. Overreactive nervous systems in animal babies keep them from exploring new environments and hinder learning. Some of the animal babies show delayed motor development and less capacity to adjust to new conditions because they are quick to overreact and slow to recover from stressors like changes in their cagemates or exposure to new social groupings. As adults, these animals have higher cortisol levels and a reduced capacity to self-regulate their emotions.

Lessened interest in exploring new situations has particularly worrisome implications for humans because such interest is fundamental to formal and informal educational success. Hyper- or hypo-aroused babies of any species are not free to focus on the world outside of themselves. These little monkeys are the counterparts to the thousands of kindergarten children who enter school each year unable to focus on learning because, with their nervous systems on "red alert," they are mounting an internal battle to try to calm themselves in what they experience as a threatening environment.

Rat pups born to mothers stressed by exposure to a loud buzzer during pregnancy, when compared with pups born to nonstressed mothers, are more fearful and irritable; they produce more stress hormones and are less able to rebalance their stress responses. Studies at the Harlow Labs at the University of Wisconsin found that monkeys born to mothers stressed during pregnancy (mostly by loud and unpredictable blasts of sound) had lower birth weights, compromised physical growth, retarded motor development and shorter attention spans than those born to nonstressed mothers.[12] They also showed increased hormonal responses to stressful events and were both more aggressive and less interested in playing with their cagemates. Other monkey babies of stressed mothers have shown a wide range of difficulties, including diminished cognitive skills, more mutual clinging, social withdrawal, attention problems, and even neuromotor impairments such as low muscle tone, poorer coordination, and a delay in learning to self-feed.[13] Extreme stress experienced by pregnant rats or monkeys is likely to result in spontaneous abortion. It appears, particularly in the monkey studies, that symptoms were more severe when stress occurred early in pregnancy.[14]

Researchers believe that the mother's HPA axis is a major influence on the HPA axis of the offspring. A dysregulated maternal nervous system may well build the same into her babies. Unless the stressors are discontinued or the babies are raised by a less upset mother—and in some cases in spite of such efforts—this pattern will continue after birth. The exact physiology of how this works is unknown, but it appears that maternal cortisol is the common denominator. Dr. Thomas Fleming and his associates who work with Dr. Barker in Southampton have observed that in sheep, maternal cortisol is directly transported across the placenta and enters the fetal system through the exchange of blood.[15] Other researchers question the assumption that maternal cortisol works this way in humans. Regardless of the precise mechanism whereby cortisol affects the fetus, variances in prenatal maternal cortisol levels appear to be a major key to variances in cortisol levels of the offspring of both animals and humans.

When the homeostatic state of the fetal HPA is disrupted by maternal stress, there are huge implications for the child's immune and endocrine health. Increased levels of cortisol in response to stress suppress immune function in both rats and primates and affect both antibody production and inflammatory responses. The effects of stressors appear to vary with the stage of pregnancy when they occur. Stress later in the pregnancy may result in suppressed immune responses in offspring, predisposing them as adults to infectious and autoimmune diseases.[16]

PRENATAL STRESS AND HUMAN BABIES

So how does all of this apply to human babies? As difficult as it is to imagine mother rats and monkeys as subjects of such studies, it is obviously harder to conduct research on human subjects, and the ethics of human subjection to such conditions makes replication of the animal studies impossible. When we were examining the roots of aggression in earliest development while researching *Ghosts from the Nursery* in the mid-1990s, we discovered a dearth of information about the impact of maternal stress on the human fetus. This was particularly troubling because, anecdotally, we heard and could easily observe correlations between anxious, chronically stressed and/or depressed mothers and anxious and overstressed babies. Even though we ourselves could *see* the connections behaviorally, no such view was available in the research. Now, with the shift from looking exclusively at the brain to looking at the relationship between the brain and the immune and endocrine systems, this research is exploding, and the results confirm our observations.

Clearly some degree of stress during pregnancy goes with the territory. It is normal, not pathological, for women to experience some stress relative to pregnancy alone—to say nothing of the accompanying realities faced by most pregnant women, including meeting the ongoing needs of older children, navigating a relationship with a partner, economic pressures, and the complexities of earning a living. But for far too many, excessive chronic stress during pregnancy takes a toll on both mother and fetus.

As we look in this chapter at the biological impact of chronic stress and trauma experienced by a pregnant woman on her unborn child, the last thing we want to do is further stress any reader who is, has been, or will be pregnant. Nonetheless, if you are in one of these categories, it is wise to do all you can to gentle your days. It has never been more critical for you to get adequate sleep, take time to relax, and maintain supportive relationships. Some women mistakenly believe that overlooking their own needs is a gift of love and what's best for their families. Pregnant or not, such a belief bodes poorly for families. Especially during pregnancy, self-care and baby care are synonymous. Your best gift to your baby is to take care of yourself and surround yourself with people who honor that priority.

For men, the message is that everything you do to support and protect your partner is an investment in your baby's health. The sheer volume of the research on prenatal maternal stress will allow you to begin to see the physical impact of this investment in the biological relationship between a mother and her baby in the womb.

Historically, our view of a fetus has been limited to observations of animals or aborted human fetuses. Recently the convergence of four advances—ultrasound imaging, neonatal intensive care unit technologies to preserve the lives of younger and younger preemies, discernment of the biology of the stress response, and new methods to evaluate stress in prenates—has given us an unprecedented look into a previously unseen world. Scientists can now observe the proliferation of cells beginning during fertilization as gametes from each parent become a zygote, then a blastocyst, an embryo, a fetus, and finally a neonate. A look inside the womb finds the fetus rehearsing for life outside. Although still considered reflexive, movements at this stage of development are surprisingly similar to behaviors after birth—grasping, blinking, sucking, head lifting, hiccupping and kicking, even defensive responses like recoiling from an amniocentesis needle—are all under way. As brain development progresses, these movements become more complex. Researchers observing these behaviors remind us that birth marks not the beginning of life but rather the change in respiration from fluid to oxygen and more complex stimulation for brain development; birth is only one developmental step in a series of many, some of which have already taken place. Dr. Nathanielsz takes this a step further: "We pass more developmental milestones during life before birth than we do after birth."[17] Certainly the steps we take before our feet can support us appear to be far more influential than we have ever before known, and more important than medical practice typically recognizes.

WOMB WINDOWS

What does this look like? Who are we prenatally, still swimming in the primordial waters of the womb? Less than three weeks after we are conceived, neurons—tiny new brain cells—are proliferating in our brains at an unimaginable rate, averaging more than 250,000 a minute. At this rate, according to Dr. Richard Restak, author of *The Infant Mind*, it would take an electrician 30 million years to finish soldering a circuit at the rate of one per second![18]

We feel touch first. By seventeen weeks gestation we can feel touch all over our bodies except for the backs and tops of our heads; sensation is felt there at twenty weeks. By the time we have gestated just two months we kick and jerk in response to being poked, and by four months we frown, squint or grimace if our scalp is tickled. If our mother drinks cold water, we kick for all we're worth. When we are ten to fifteen weeks along, a cough from mother causes us to move rapidly. And if she is terrified, we feel her chemistry. We kick, and our hearts beat extra hard.

By fifteen weeks we can taste differences in the amniotic fluid because of the presence of varying chemicals. By twenty weeks we can hear
the voices of our mother, our father, our siblings and others who are
often close by. After we are born, we recognize our mother's and father's
voices if father has been close to mother regularly, and we can distinguish their voices from others. We show our preferences by turning our
heads toward familiar voices and away from unfamiliar ones. By twenty-
four weeks the sounds of Vivaldi and Bach and Mozart calm us if we
are lucky enough to hear them. We recognize this music after birth, as
well as any other songs or nursery rhymes we have heard for the last
trimester. We demonstrate our familiarity with these rhymes by sucking in a rhythm that matches their cadence. We don't do this with
rhymes we don't recognize.

At just sixteen weeks we move away from a bright light held just outside our mother's belly. We may cover our eyes with our hands, just as we
cover our ears when we hear the ultrasound. At thirty-two weeks we can
see as well as we can at thirty-eight weeks when we are born. By twenty-
six weeks, if mother is very upset, we also feel upset and unsettled, actively moving our limbs and whole bodies. If this goes on, we may be
born too soon and will not be ready.

If born prematurely, we are in excruciating pain because we have no
insulation; we may feel the pain of our expulsion from the womb even
before we are fully born. Sound, light, touch and texture cause distress.
Our eyes and ears and skin are not ready. We can't cope yet; we are in
trauma. Development as we know it has stopped and been replaced by
constant overload and pain. We sleep and waken to more of the same,
but we seem more stilled if we're in contact with our mother's skin, where
the smells and the rhythms and the sounds are familiar. We need a quiet,
darkened, cozy, warm, moist environment.

TEA FOR TWO

The use of ultrasound and sonograms allows an exceptionally early window on the effect of maternal stress on learning and temperament.
Through ultrasound, researchers can observe that high maternal cortisol
may begin to impinge on learning in the womb. Tests of fetal maturity

measure heart rate in response to sound and light. After a certain number of successive sounds or light exposures, the expectation is that, as the fetus matures, she or he will show habituation or recognition of the pattern. Heart rate—which rises initially when sound or light is presented—will gradually "habituate" so that the fetus is no longer aroused. This is the same pattern we see in a healthy infant when asleep or in a quiet alert state. If we shake a rattle just a few feet away, a healthy baby initially startles, then after a few times the infant subconsciously recognizes the sound, and his or her heart rate no longer rises in response—a sign of "habituation."

Studies have shown that fetuses of mothers with chronically elevated cortisol levels are less able to distinguish repeated tones from novel ones, showing little variation in their heart rate between new sounds and the repetition of what should be familiar sounds. Since habituation, as measured by heart rate, is predictive of cognitive development, this outcome is of real concern. How long the effect endures is not yet known.[19] In one study, Dr. Pathik Wadhwa, a researcher at the University of California, tested fetal habituation by exposing them to new noises and sounds (vibrations). When the mother was not stressed, the fetus, though initially startled, settled down fairly quickly. If the mother was stressed, not only did the fetus react more strongly and for a longer time, but these sources of extra stimulation also tripled the risk of premature birth.[20]

When pregnant women are anxious or depressed, their fetuses reflect all the signs of stress, particularly in the last trimester: the physiology and behavior of these fetuses are dysregulated in very much the same way that is seen in traumatized NICU preemies.[21] Dr. Paula Thomson, a Los Angeles psychologist specializing in attachment, has synthesized more than one hundred studies that make a strong case that the prenate, while lacking the cognitive capacity to interpret events as traumatic, nonetheless experiences his or her own form of trauma through the mother's chemistry. Dr. Thomson believes that the chemistry of prenatal experiences rivals genetics in the transmission of generational trauma and that much of what is currently viewed as "genetic" is actually "congenital"—attributable to environmental influences in the womb. Thomson asserts that it is crucial to identify high levels of stress in pregnant women in order to attempt to interrupt a cycle that can reverberate for generations. Medical professionals typically interpret the symptoms of trauma in newborn behaviors as acute responses to birth or to experiences immediately following. In fact, Thomson warns, much of what we are currently passing off as resulting from neonatal experience is actually a reflection of life in the womb.[22]

Mothers across the world have long reported that a sudden experience of fear or anger increases the quantity and intensity of fetal movements during their second and third trimesters. Some researchers have found that mild prenatal stress can actually give the fetus an advantage in learning, in respiratory maturity, and even reduced emotionality, especially during the last trimester. No one disputes, however, that chronic stress presents a very different picture. Unfortunately, chronic stress is increasingly the daily reality for families throughout the developing world, as well as for those experiencing lower socioeconomic conditions in the Western world, where the battle to afford basic nutrition and housing— to say nothing of medical care—is often waged with little social support. When challenges within a mother's external world are added to an unresolved history of early relational trauma or loss, current domestic violence or mental illness, her system may transfer the biology of chronic stress or trauma to another generation. If these conditions persist postnatally, the combination takes a huge toll on the physical foundations for the endocrine, immune and cardiovascular processes essential for stable emotional regulation and the health of the next generation.

Signs of stress in the fetus are similar to stress responses in infants: elevated heart rate, increased motor activity, "pronounced stilling," grimacing and dysregulated sleep-wake states—essentially disturbances in both physiology and behavior. Thomson, like Fleming, points to the placenta as the nexus of exchange between the mother and baby, "a shared region where the mother's and prenate's blood supply interact in an ever changing amplifying or dampening process."[23] Some researchers believe that the formation of stable or unstable regulation of the HPA axis is determined in the womb. Nathanielsz calls this "programming"—a process whereby early conditions shape a lifelong biological function.[24] Essentially, stressors that cause anxiety (including hunger) or depression in the mother trigger her HPA axis to circulate high levels of cortisol in her body.[25] Enzymes in the placenta usually inhibit excess maternal cortisol from interfering in fetal growth, inactivating it as it attempts to cross.

But this protection is often incomplete. Placentas that are small or not functioning properly may lack sufficient protective enzymes to shield the fetus from the effects of excessive cortisol. This is especially true when the mother is highly stressed. Many maternal health factors can lead to a small or poorly functioning placenta. Add in stress and her fetus is endangered—often in subtle ways that we don't notice; increased risk for heart disease, diabetes and obesity or an HPA axis that is unstable and hypersensitive to later stressors may begin here. Nathanielsz says, "It is when the mother is subjected to an overwhelming set of adverse circumstances—poor nutrition, constant financial worry, physical abuse, drugs, alcohol and other extreme circumstances—that the placental barrier may prove inadequate."[26]

To further compound this cycle for highly stressed mothers, during the last weeks of pregnancy the now-affected placenta may begin to fail in its task of transporting oxygen and glucose to the fetus, and the fetus responds by secreting cortisol of its own. The combination of excess cortisol resulting from mother's stress passing the placental barrier and the cortisol being secreted by the fetus itself is doubly harmful, causing extreme sensitization of the fetus to stress and trauma. Dr. Robert Sapolsky of Stanford University calls this process and the resulting extrasensitivity to stress "endangerment." His work on stress in animals and humans shows that lifelong overexposure to cortisol causes rapid aging and premature death in rats and in humans. Recall that excess cortisol is also the chemical culprit behind damage to the hippocampus in PTSD. Dr. Wadhwa explains: "When the mother is stressed, several biological changes occur, including elevation of stress hormones and increased likelihood of intrauterine infection. . . . The fetus builds itself permanently to deal with this kind of high stress environment and once it's born may be at greatest risk for a whole bunch of stress related pathologies."[27] Too often, such babies emerge with their little fists already clenched, ready to fight for their lives in a dangerous world. And their capacity for higher thinking and relating may be undermined by a perceived need to stay focused on survival. The studies have tremendous implications for the need to adequately discern and gentle pregnant women's emotional states and to intervene prenatally and perinatally, when it makes the most difference both for her and for her baby.

Studies of children born to mothers who were pregnant during the course of disasters such as the Holocaust and 9/11 show the impact in particular of PTSD. A study of thirty-eight babies born to pregnant women who were at or near the World Trade Center that morning gives a very clear view of how the youngest survivors fared on a day etched into all of our memories.[28] Measuring cortisol levels in the saliva of the pregnant women, researchers looked for signs of stress, particularly for low cortisol levels, a cardinal marker of PTSD. Mothers who developed anxiety and other diagnosable symptoms in response to their trauma had significantly lower rates of cortisol than mothers with minimal symptoms. Lowered cortisol levels in the babies matched those of their mothers in infancy and again at one year. Babies of mothers in their last trimester of pregnancy at the time of the attack showed the most pronounced effects. Lead researcher Dr. Rachel Yehuda, a psychologist specializing in combat-related distress disorders, said, "The findings suggest that mechanisms for trans-generational transmission of trauma may have to do with very early parent-child attachments, and possibly even in utero effects."[29]

The adult children of mothers who survived the Holocaust show similar patterns to the 9/11 babies. Historically, we have believed that the lowered cortisol levels of the children of Holocaust survivors were attributable to environmental factors relatively later in development, such as living with anxious, fearful or depressed mothers, or to vicarious trauma from listening to their parents' stories. Shortly after her assessment of lowered cortisol levels in the 9/11 infants, Yehuda revisited a population of adult children whose mothers were pregnant during the Holocaust. Based on questionnaires, she divided them into two groups: those whose parents had PTSD and those whose parents did not. The adult children's blood cortisol levels were measured every thirty minutes for twenty-four hours. Yehuda found that those children whose mothers had PTSD had lower cortisol levels—levels similar to those of trauma survivors with PTSD—*even though none of the adult children had experienced the Holocaust or had other symptoms of PTSD.* She concludes that cortisol programming may in fact take place in the womb through an epigenetic effect wherein the expression of genes is driven by chemical messages.[30]

SURVIVAL TACTICS

To recognize the profound toll from maternal stress or malnutrition across species is to realize that there must be an evolutionary advantage at play. For any organism, survival and reproduction are the fundamental goals; processes that shape development are driven by these goals. Historically we have believed that babies develop in the womb on an automatic course shaped almost entirely by genetics. Now this thinking has been turned upside down.

Dr. Peter Gluckman, professor of pediatrics at the University of Auckland, and Dr. Mark Hanson of the University of Southampton, in their book *Mismatch*, explain the pervasive behavioral and learning differences in the offspring of prenatally stressed mothers, citing studies of snowshoe hares in North America, a species that declines in population when food is scarce. With fewer hares overall, the survivors are more likely to be killed off by their natural predators. Thus, to preserve the species, the remaining hares must be exceptionally vigilant. Stressed female hares, fearing for their lives, inadvertently wire their offspring accordingly and give birth to more vigilant babies, increasing the likelihood that the offspring will survive the scarce food supply until they can reproduce. Gluckman and Hanson call this process "predictive adaptive response." However, while hypervigilance and the capacity for rapid shifts of attention are adaptive for a dangerous predator-ridden environment, they may be maladaptive in a normal environment.[31]

Genes provide a flexible blueprint for the fetus, a potential framework. Genes are constantly interacting with the environment in the womb, taking their cues from mother's chemistry to adjust fetal development to the world that mother's body is communicating hormonally.[32] Rather than proceeding on a course set entirely by DNA, genes are constantly "reading" the environment in the womb for cues about the world outside to gear the fetus for survival. If mother is stressed and her HPA axis dysregulated, her baby's is likely to be also.

For women, the state of pregnancy itself adds unique additional stressors. Alterations in appearance, hormonal changes affecting mood, and anxieties specific to pregnancy, like fear of pain in childbirth and fear of an abnormality in the baby, may be added to a pregnant woman's worries. A Dutch study of more than two hundred mothers used three different assessment tools to measure the impact of pregnancy-related stress on their babies. One was a measure of "daily hassles," filled out by the mothers. The second tool measured pregnancy-specific anxiety: mothers rated their responses to ten common concerns, including fear of giving birth, fear of bearing a physically or mentally handicapped child, and changes in their appearance. The third assessment was of the mother's own perceived levels of stress. In addition, maternal cortisol samples were taken to assess stress regularly during each of the three trimesters of pregnancy. Babies' temperaments and cognitive development were evaluated at three and eight months after birth, and the results were correlated with the mother's prenatal stress scores. The Dutch researchers found that maternal stress, particularly in the first half of pregnancy, had a negative effect on babies' temperaments at both three and eight months. Pregnancy-specific fears such as worries about the baby's health or about pain during delivery seemed to have the strongest correlation to negative outcomes in the babies' behavior and temperament at eight months; babies' capacity to regulate attention appeared to be most affected. Mothers' levels of perceived stress during pregnancy particularly correlated with their babies having more problems adapting to new situations and to increased distress around unfamiliar people.[33]

A second Dutch study of 170 first-time mothers found that high levels of stress and anxiety, especially pregnancy-specific anxieties in the first trimester of pregnancy, were associated with infants' lowered developmental scores on the Bayley Scales of Infant Development, a standardized tool for assessing child development. Babies had more difficulty adapting to a new environment and more problem behavior at age six months. A Belgian study followed seventy mother-infant pairs from the first trimester of pregnancy to age nine. Infants of highly anxious women had more difficult temperaments from thirty-six weeks gestational age and throughout the first seven months after birth. They cried more, were more physically active, and were more irritable. They also had more irregular biological functions (eating, sleeping and eliminating). At age nine, these children, especially the boys, were still more active, had a higher rate of attention deficits, and were more aggressive and impulsive.[34]

A recent study in London focused on mothers who came to an amniocentesis clinic with concerns about possible Down syndrome in their fetuses. All of the women in the study had normal fetuses and ultimately gave birth to normal, full-term babies. The researchers collected a vast amount of information on the mothers and babies over the first fourteen to nineteen months of the babies' lives, including thorough histories of the mothers. Prenatal stress in mothers—in this case, stress concerning the health of their babies—was found to be predictive of both increased fearfulness and compromised mental development in their babies, but none of the babies exhibited both effects: it was an either-or result. The researchers speculated that these differences might have had to do with the timing of the stressor or that genetic factors also played a role.[35] Of particular note is that the relationship between the mother and her partner emerged as most predictive of both the emotional and cognitive problems faced by babies at fourteen to nineteen months, pointing to a critical need for identifying and intervening in domestic violence early in pregnancy.

It would be hard to find many of us who are not stressed at least somewhat, especially during a first pregnancy. The question is, how much stress is manageable? This has been the subject of several studies. High levels of maternal stress threaten not just the quality of life but life itself: the baby may not survive. Preterm birth and low birth weight have long been associated with maternal stress. Again, the linchpin is cortisol. At the end of pregnancy, it is natural for mothers' cortisol levels to rise in preparation for birth. But high levels earlier in the pregnancy often trigger early labor, reduced head circumference, and other malformations associated with increased risk of developmental delay, learning disorders, chronic lung disease—and even death.

Cortisol receptors are plentiful in mothers' reproductive systems. The endometrium and the ovaries, for example, contain abundant receptors for cortisol and corticotropin-releasing hormone (CRH), so it is not surprising that stress can have such a deleterious impact on sexual and reproductive capacities. Most of us are aware that chronic stress dulls our desire for sex. And we are all familiar with the couple who has tried and tried to conceive a baby, and when they finally decide to adopt the woman quickly becomes pregnant. A study of in vitro patients found that those with a high degree of stress or anxiety had a greater risk of not carrying the baby to term. In humans there is a direct correlation between prenatal maternal stress and several complications of pregnancy, including an increased risk of spontaneous abortion and of developing pre-eclampsia. Currently one in ten women in the United States delivers preterm—before thirty-seven weeks.[36]

In epidemiological studies, extreme maternal stress during pregnancy is also linked to schizophrenia, autism and vulnerability to anxiety and depression in the babies. A recent report from the New York University School of Medicine showed high rates of schizophrenia in the children of mothers who were pregnant during the Arab-Israeli War. This study correlated the birth and health records of almost 89,000 people born in Jerusalem between 1964 and 1976. It found that people with schizophrenia were more than twice as likely to have been born in January 1968; their mothers would have been in their second month of pregnancy during the Six-Day War in June 1967. The risk of schizophrenia was four times greater in females than in males. The risk for other psychiatric disorders was also two and a half times higher for children whose mothers were in their third month of pregnancy in June 1967. Principal investigator Dr. Dolores Malaspina says:

> Our study added to the strong and accumulating evidence that maternal psychological stress alone (without infection, malnutrition or other prenatal adversity) can have lasting effects on the health of the offspring. These changes may reflect an evolutionary benefit of adjusting the behavior or metabolism of a baby that is born during a stressful period. Clearly schizophrenia is not adaptive but perhaps prenatal stress is acting to make someone more sensitive to potential threats or more vigilant. These might be the intended effects that increase the risk for schizophrenia.[37]

Confirming Malaspina's findings, a Danish study of 1.38 million births from 1973 to 1995 reported that loss of a close relative during the first trimester (but not during the second or third trimesters) significantly increased the risk for schizophrenia in the offspring, even with no family history of schizophrenia.[38]

Similar studies have found that childhood behavior in general is often compromised—socially, emotionally and academically—by high rates of maternal prenatal stress. A 2002 study of more than seven thousand mothers and babies from the Avon region of England found that highly stressed women were twice as likely as nonstressed women to have children with behavioral challenges, depression and anxiety when the children were ages four to seven.[39] Effects may show up at birth or shortly thereafter—though most of us are unsophisticated about the early signs and what to do about them. Babies of mothers who were highly stressed during gestation are fussier and react more negatively to novel people and situations.[40] Researchers believe that this may indicate that differences in temperament emerge *before* birth. A behavioral profile of high reactivity, negative affect and motor activity when a baby is introduced to new toys has been associated with later shyness, anxiety disorders and language delays.

Prenatal stress also generally correlates with higher rates of social and emotional problems during childhood and with higher rates of ADHD observed in children ages four to fifteen.[41] In a study comparing the pregnancy histories of 188 women who delivered autistic children with those of 212 women who had normal children, Dr. Davis Beaversdorf, a neurologist at Ohio State University Medical Center, found that the stress levels for mothers of autistic children were twice the levels of other mothers, including mothers who delivered Down syndrome children. Mothers of autistic children were likely to have experienced a major stressor such as the death of a spouse, a job loss, or a long-distance move midway through their pregnancies between the twenty-fourth and twenty-eighth weeks. Beaversdorf believes that the prenatal environment, particularly the chemistry of maternal stress during specific periods in pregnancy, may lead to structural changes in the brain, especially the cerebellum, a part of the brain that is altered in autistic children. Although he agrees that there may be a genetic component also, his theory is that it is the interaction of the two factors that leads to autism. Usually undetected at birth, autism tends to appear before age three.[42]

PRENATAL MATERNAL DEPRESSION

Dr. Tiffany Field, director of the Touch Research Institute in Miami, has been a tireless advocate for the importance of touch and massage in reducing stress for mothers and infants. Field's research shows that fetuses and infants of prenatally depressed women are little mirrors of their mother's stress—physiologically, behaviorally and biochemically. Peeking in through ultrasound, Field has found that the fetus of a depressed mother reflects her stress via greater activity levels, an elevated heart rate and growth retardation. Like the babies of anxious mothers, these fetuses also have a much greater risk of dying: they are more likely to be born early, to abort spontaneously, or to suffer placental abnormalities. At birth they are at greater risk for bronchopulmonary disease, intraventricular hemorrhage and being small for gestational age; they are often more irritable and less active. As infants, they often have elevated cortisol levels and lower dopamine and serotonin levels, resulting in less positive affect. They have more difficulty with orienting to the assessment process (for example, the Brazelton Inventory) and adapting to new stimuli and people. These effects may continue at least through toddlerhood. Field estimates that 10 to 25 percent of pregnant women suffer from depression.[43]

Alcohol consumption adds a particularly nefarious element to the impact of maternal stress on babies. Babies born to women who are using alcohol at the time of conception reflect some of the same challenges as those born to women stressed during their pregnancies. A Toronto study of fifty-five babies born to women in an alcohol intervention study showed elevated stress hormones, elevated heart rates and negative affect at ages five to seven months compared to babies born to nondrinking women. Although the mothers abstained after learning they were pregnant, researchers believe that the effects, similar to results of the research on undernutrition, were set in motion between conception and pregnancy recognition (typically about six weeks after conception). Not only the infant's response to stress but his or her ability to recover from stress was impaired, and boys were even more affected than girls. We have known for some time about the effects of fetal alcohol syndrome (FAS) and fetal alcohol effects (FAE), including impulsivity, attention and judgment problems, physical deformities and retardation. But this study, which evaluates subtle evidence of impaired emotional regulation, adds a dimension never before recognized regarding the timing of exposure and suggests that women intending to be pregnant are smart to avoid alcohol altogether.[44]

Although there is still much to sort out as to cause and effect and the interaction of genetic variables, the results of these early studies on humans are generally consistent with those from animal studies and have uncovered some clear and inconvenient truths.[45] Maternal stress alone, conveyed unintentionally in the secluded chamber of the womb, can exact a toll on children's health and behavior by dysregulating the chemistry of fear in the unborn. Unrecognized until very recently in the scientific community, the impact of maternal stress has yet to be addressed in our health care systems. Pointing the finger at mothers is missing the point. Mothers are not to blame and cannot shoulder the weight of this growing problem. This is a societal, community and family issue, the purview as much of fathers as of mothers. Prenatal depression is a health problem that not only has an impact on the larger community but can't be resolved without the involvement of that community. We are all affected. Yet most doctors are uncomfortable with the topic and are not trained to interview women regarding these issues (especially middle- and upper-class women); even when they do uncover chronic stress in patients' lives, most doctors don't know where to send them for help. Only the most obvious cases are likely to be diagnosed. Meanwhile, rates of anxiety and depression are soaring among women of childbearing age. The proliferation of nervous systems skewed by stress will continue until we recognize the wisdom of the ancients and are willing to pay the price of gentling life around pregnant women.

PERINATAL TRAUMA

The *perinatal period* technically includes the last six months of prenatal life, but in common usage the term refers to the period shortly before, during and just after birth. The perinatal period represents a discrete developmental period in the research, and as technology has progressed, this time of development has become far more interesting to researchers, particularly those examining the influence of this period on our evolving nervous systems.

Prematurity

Being born too soon is widely recognized as a traumatic beginning. Regrettably, this is a rising problem: one in thirteen babies in this country are now born prematurely. The percentage of infants delivered at less than thirty-seven weeks has climbed 20 percent since 1990 and 9 percent since 2000, affecting 12.6 percent of all white births and 18 percent of all African American births.[46] Of those black babies, nearly 4 percent are born before thirty-two weeks—two and a half times the rate for white babies. Prematurely born brains are traumatized brains, which put babies at risk of adverse health and mental health outcomes. Approximately 52 percent of children born prematurely develop school problems and emotional disabilities. For NICU babies, the combined stress of being isolated from the mother and the daily pain and discomfort of a NICU unit leads to what Dr. Heidelise Als of Harvard Medical School calls "neurotoxic brain-altering events"—in other words, trauma. Dr. Als is world-renowned for revolutionizing NICU care to honor the emotional needs of these tiny patients and their families.[47]

Chuck Green, president of Alpha Maxx Healthcare, which serves pregnant women and their infants for the first year of life, says simply: "NICU babies are the most vulnerable humans on earth." Green's firm serves pregnant Medicaid mothers in western Tennessee, a population at very high risk of negative birth outcomes, including preterm birth, as a consequence of substance abuse issues, smoking, domestic violence, low literacy and homelessness. He believes that racism and poverty are traumatizing multigenerational influences on these families, who deliver a high rate of NICU babies. He and his compassionate staff work valiantly and creatively with the women they serve during their pregnancies and the first months of their infants' lives, using trauma-focused techniques in an effort to reduce deleterious health outcomes in yet another generation. They have shown great success, reducing the average NICU baby's length of stay from 14 days to 9.8 days, a 31 percent reduction.[48]

In the NICU the baby is engaged in a struggle for life while enduring highly invasive procedures, including the insertion of needles, catheters and tubes into veins, arteries, lungs, bladder and stomach to facilitate essential processes like eating, breathing and extracting vital liquids to measure how well the child is doing. Babies are ensconced in enclosures designed to simulate conditions in the womb; however, these high-tech marvels, while medically necessary, cannot begin to give them what they need the most—their mother's womb and the comfort inherent in that dynamically attuned relationship. Our best technological efforts are a far cry from the environment that nature intended—a rich sensory environment exquisitely tailored to their gradual development. Within the womb, the fetus's new ears are tuned by exposure to the sounds of the mother's digestion and a chorus of placental sounds, as well as the mother's heartbeat. The fetus perceives muted sounds from outside as well—most important, mother's voice—and is rocked by the vestibular rhythm of mother's movement, insulated perfectly for the level of touch and sound that the developing nervous system can withstand. And there are the tastes of myriad chemicals in the amniotic fluid, reflecting the mother's diet. All of this is perfectly titrated preparation for the world outside— the world after birth. When the fetus is expelled too soon from this environment, the result is trauma. Fear hovers in the NICU, in spite of great effort on everyone's part, especially that of enlightened and caring nurses. Minimizing pain and trauma is what these units are all about, but just being there takes a pervasive toll on the tiny patients. In the words of one consortium of professionals administering the protocols to preemie patients:

> Neonates in the NICU are exposed to acute, repetitive, and chronic pain as a result of procedures, surgeries, and disease processes throughout their hospital course. Preterm neonates, especially those less than 30 weeks gestation, are exposed to 10–15 painful procedures per day at a time when pain is developmentally unexpected. Pain left untreated leads to changes in neurologic processing, increased excitotoxicity, and cell death. This can lead to ADHD, deficits in cognition, learning disorders, and motor abnormalities.[49]

Researchers like Taddio and Anand explain that pain neurotransmitters are functional and abundant in the fetus before birth; the somatosensory cortex already contains "large receptive fields of neurons."[50] The number of pain-receptive nerve fibers in the skin of a neonate is similar to, and possibly even greater than, the number in an adult. So, during the perinatal period, not only is the fetus highly sensitive to pain, but the counterbalancing cortical mechanisms in the fetal and infant brain that later help modulate pain have not yet developed, rendering the fetus and the newborn relatively less able to cope with pain. Although painful procedures may be essential to preserving these fragile lives, and none of us would deny these amazing procedures to a baby, we must recognize that there is a price to be paid in future development. Babies who begin their lives in chronic pain are at risk not only of childhood sleep problems, difficulties in self-regulation and attention and learning disorders but also of long-term alterations in pain perception, chronic pain syndromes and somatic complaints in adulthood.

Such outcomes are due to the fact that daily experiences of pain can trigger excessive *hyperinnervation*—the innervation of pain systems—which can result in lowered pain thresholds and hypersensitivity to even normal stimulation like routine handling or touch. Persistent pain experienced in early childhood creates molecular changes in the child's nervous system, resulting in a biological marker that is recorded in implicit subconscious memory in the lower brain. Such markers can become the basis for later somatic illnesses that seem to have no basis in the individual's current physiology, including fibromyalgia, chronic pain syndrome, and various stomach, joint or other random complaints that defy medical diagnostics. Very-low-birth-weight babies are also often at risk. Several studies have confirmed that these babies report more physical complaints in later childhood than children born at full term. There is also a correlation between the early experience of chronic pain and later attentional and emotional problems, probably because the anterior cingulate cortex, the part of the brain most affected by pain, is also closely connected to the parts of the brain associated with emotion and attention.[51]

Mitch and Grace are an example. Fraternal twins, now seven years old, these two had the good fortune to be born to Caroline and Garin, a professional couple with the means, education and commitment to be totally devoted to the twins when they were born. Delivered after just thirty-two weeks gestation, Mitch and Grace, though six weeks early, suffered no huge discernible crises after they were born. Because Grace had begun to show signs of intrauterine growth retardation, the babies had to be delivered early and were given two days of steroid injections to ready their organs, particularly their lungs, for life outside the womb. Born just a minute apart by cesarean section, Mitch required some extra help with respiration and was put in a separate isolette. Grace spent three or four days under the bilirubin lights to help alleviate her jaundice. But after the initial few days the two were reunited in an isolette, where Mitch continued his prebirth habit of sucking on Grace's head as they curled into each other, just as they had in the womb. Caroline visited the NICU for the better part of each day for the week she remained in the hospital. Subsequent to her release, she came to the hospital to be with her babies for seven or eight hours daily. Garin also came "just to hold them" as regularly as he could. For almost a month, their daily ritual revolved around being with the babies in the NICU. For Caroline and Garin, the staff became, as they had been for many others, an extended family. Caroline pumped and carried breast milk from home and spent her days feeding and holding both babies on her chest while she rocked them "skin to skin": "For most of those days I just sat in a rocking chair with one or both of them on top of me," Caroline said.[52]

When asked about any painful procedures that her babies were subjected to, Caroline, who had clearly been in the throes of her own trauma at the time, said that she didn't remember any such procedures, "except I know they had a lot of blood work." She recalled the wires and tubes and how fearful she was of undoing one of them as she handled her babies. And she remembered the sounds, the frequent alarms and the cries of other infants—especially the screams of one little one who cried for most of twenty-four hours from hunger because of a digestive malformation. Caroline, like most grateful parents for whom NICU nurses literally hold their hearts in their hands, was effusively grateful about the quality of nursing her babies received. Mitch and Grace came home just over a month after they were born.

Both children progressed well through developmental milestones—crawling, walking, talking—within a normal time frame. But for the first four or five years of their lives each child wore a startled look. Their eyes were open wide, as if they had been suddenly surprised; a "deer in the headlights" look—what some might say is the look of trauma—is only now diminishing. As is common with preemies, and also for twins or multiple-birth children, both are immature for their chronological age. While their peers in kindergarten were initially further along in their letters, numbers and writing, the twins are obviously bright and progressing at a steady rate. By first grade, they caught up with their classmates, especially Mitch, who delights in reading *Captain Underpants* to anyone who will listen. As a toddler, Grace was shy and slow to warm to new people and places, but that shifted by first grade, and now she often has to be reminded about "not talking to your neighbors" in her classroom. Mitch has until very recently been prone to being whiny, ritualistic and easily frustrated. If, for example, a toy wouldn't do what he wanted, he became explosively angry, often destroying it in his frustration. It was hard to distract and redirect his attention, so Mitch became known for long bouts of negative emotion. Although both twins tend to take their frustration out on each other, they also consistently seek each other out as favored playmates and are as inclusive of the other as they are competitive. Mitch is still painfully sensitive to odors; he becomes upset beyond comfort if exposed to a bad smell, including those associated with the family dog and his own body.

Interestingly, when asked what, if anything, she sees in Mitch's and Grace's current level of development that seems unusual, Caroline's first comment is that Mitch and Grace both overrespond to "oww-ees." She notes that "a tiny scrape or scratch will set off a big reaction that takes a giant kiss and a big production to calm down. . . . It's really hard to distract them from even a tiny hurt." Here is the echo of earliest trauma. Without knowing it, Mitch and Grace each reflect the imprint of their painful beginnings. With time, this hypersensitivity to pain may ebb. But it also may resurface in another form later, disconnected from its source.

The correlation between prematurity and compromised health outcomes is not surprising. While death may be avoided in babies born at less than thirty-two weeks gestation, they often suffer lifelong health consequences, including cerebral palsy, mental retardation and chronic lung disease as well as visual and hearing problems. The National Research Center for Growth and Development of New Zealand estimated in 2008 that heart disease and diabetes would be reduced by 44 percent and 66 percent, respectively, if both fetal and infant growth could be optimized. Like babies who are small for their gestational age and low-birth-weight babies, premature babies are more prone to type 2 diabetes, hypertension, coronary artery disease and stroke in adulthood. Low birth weight has also been correlated with abnormally high stress responses later, particularly in men.[53] The research indicates that insulin sensitivity is affected in these babies, and they are also at higher risk of developing metabolic syndrome, which includes the cluster of medical disorders that greatly increase the risk of cardiovascular disease: high blood pressure, high blood sugar, high cholesterol and a higher ratio of fat around the waist.

Birth

Even for those babies fortunate enough to complete all forty weeks of their intended gestation in a well-nourished and minimally stressed prenatal environment, birth is a huge developmental transition from their previously passive experience in the womb. Even when they experience no complications of any sort, the unbuffered exposure to visual, tactile, olfactory and audial sensations outside of the womb presents an almost unimaginable adjustment, overwhelming in its impact. While some therapists talk of "birth trauma," there is still debate about the validity of this concept. Many assert that even in difficult births, if labor has begun and proceeded for a while on its own, there are natural, chemical changes in both the mother's and baby's systems, such as heightened endorphins and oxytocin, that are stimulated by the birthing process and provide an analgesic and calming effect.

But various common obstetrical practices—the use of forceps and vacuum, some forms of anesthesia or drugs given to the mother for pain, or induction of labor—may lead to trauma for the infant during birth. Clearly there are times when these interventions are life-saving and entirely necessary. But unless there is a life-threatening reason to use them, some practices, particularly inductions for superficial reasons, should be avoided.

Labor induction, practiced for medical convenience and often to match a family's plans for vacation, job return, and so on, has risen in popularity, more than doubling in frequency since 1990 to 22.3 percent of all births.[54] According to several experts, the short-term benefit of rushing a birth is nothing compared to the long-term consequences for the child, which can include lifelong issues with motivation, quitting, punctuality, and a resistance to being hurried.[55]

There is also a voluminous body of research showing that protracted labor—a baby spending long hours in the birth canal, having the umbilical cord wrapped around his or her neck, being in a breech or transverse position, or having to be forcibly manipulated to accomplish delivery—may become a somatically registered memory that resonates in later fears. In their fascinating book *Trauma Through a Child's Eyes*, Peter Levine and Maggie Kline tell the story of "Remarkable Devin," an eight-year-old who presented with several diagnoses, including ADHD, auditory processing deficit, sensory integration dysfunction and dyspraxia. He was hyperactive and aggressive, and he had focusing problems and trouble completing assignments in school. Devin was a daydreamer (a term often used by teachers to describe a dissociative child) and had terrifying nightmares. His mother's description of her pregnancy and Devin's birth shed some light on the problems. During the twenty-first week of her pregnancy, she was medicated to prevent premature labor and placed on bed rest. At thirty-six weeks, she stopped the medication and went into a weak labor that the doctors decided to boost with pitocin. In severe pain, she was given a narcotic. The baby was in distress for two hours before doctors discovered that the umbilical cord was wrapped around his neck three times. They used suction to deliver Devin, who was blue at birth. At age three weeks, he was hospitalized with jaundice and anemia, and the weekly blood draws required staff to hold him down screaming. Devin's was a traumatic start by any standard.

Devin began his play therapy by drawing a rocket with a rocket launcher and trailer, explaining: "The astronaut was glued inside, and it would take ten seat belts to make it safe, and he would have to wear two astronaut suits to come out because it's so cold out there." Astonished, the therapist watched as Devin added a lifeline to the astronaut and connected it to the rocket, saying: "The cord is the most important part!" After several sessions in which the therapist validated Devin's anger, Devin began to use materials in the office to perfectly reconstruct his birth trauma. He made a tunnel of pillows, then reached out and grabbed a nearby electrical cord and wrapped it around his neck. Gently the therapist assisted him in unwrapping the cord and continuing on his journey through the tunnel. After writhing and squeezing through at a pace all his own, Devin crawled over to hug his mother, who was waiting there to embrace him. His breathing deepened and calmed. This was a transformational event in the life of the little boy. His tantrums diminished, and he began to persevere and complete his work. Devin still wears some signs of trauma and will continue to need support to learn to moderate sensations of terror in which he holds his breath and tightens his belly.[56]

Many clinicians believe that claustrophobia and other phobias like fear of the dark are rooted in prolonged containment in the birth canal and are unconscious memories buried deep in the limbic brain. Dr. Robert Scaer believes that the activity of the brain stem is all that is needed for some of these experiences to be recorded in unconscious or procedural memory: "Initiation of survival-based behavior requires an immediate unconscious response, not conscious planning. It is very likely, therefore, that infants process a great deal of information through mechanisms involving procedural memory and begin to assemble their repertoire of survival-based learning long before conscious memory is developed through myelination of hippocampal and cortical pathways."[57]

Medicalization of birth has led to several practices that may be adding to the rate of trauma and subsequent pathologies in newborns: not only birth inductions but cesarean births, amniocentesis and the use of various forms of anesthesia, including epidurals, which slow down labor. This is not to cast blame on anyone. But it is to say that we need to be aware, cautious and willing to be our own well-informed advocates. We live in a litigious society in which many of us expect that our medical institutions will provide a risk-free environment for birth using all possible technologies and interventions. Some fearfulness is natural as we face giving birth. But living as we do in a fear-saturated culture, it takes unusual mindfulness to resist letting this searing emotion become a force in our birthing choices.

Obstetrics has seen a progressive increase in practices that are designed both to ensure the birth of an undamaged baby and to protect against litigation. World Health Organization (WHO) data indicate that both mothers and babies generally do best where cesarean section rates are between 5 and 10 percent and that rates above 15 percent do more harm than good. In 1965 the C-section rate in the United States was 4.5 percent. By 2007 it had risen to 32.8 percent. There are two popular explanations, neither of which appears to be accurate. First, many health professionals and journalists point a finger at mothers, saying that they are asking for C-sections for nonmedical reasons. However, a "Listening to Mothers" survey done by Childbirth Connection with mothers who gave birth in U.S. hospitals in 2005 found that just one woman among the 1,600 participants said that she had a planned C-section at her own request and for no medical reason. Surveys in other countries report similar results. A second common explanation for higher C-section rates is that the number of older women giving birth is increasing, leading to more medical complications and higher rates of multiple births. The facts, however, also do not support this explanation. Despite the overall changes in the population of childbearing women, C-section rates are increasing for all women regardless of their age, the number of babies they have had, the extent of their health problems, or their race or ethnicity.

There appear to be several reasons for the increase in C-sections. First, enhancing a mother's ability to give birth naturally is often a low priority. Many hospitals discourage the use of a doula or low-tech interventions like hand-to-belly movements to turn a breech baby. Also contributing to increased C-sections are interventions in labor, such as an induction in a first-time mother when her cervix is not soft and ready. Yet another factor is that physicians discourage women from having a vaginal delivery after a cesarean. Today nine out of ten women with a prior C-section are having cesareans in subsequent births.[58] Americans seem to have a very limited awareness of the dangers associated with the C-section, which is a major surgical procedure with all the attendant risks. Both fear of litigation and lack of adequate compensation for the extra time and attention that may be needed to support a vaginal birth are cited by the survey as additional factors. And C-sections are much more efficient in terms of office planning, personal life and hospital workload, and they may offer doctors and hospitals a much greater opportunity for profit.[59]

Collectively, we who are unaware of our own trauma are particularly vulnerable at times like childbirth and may, driven by our unrecognized fears, inadvertently re-create the same for the next generation. In a Norwegian study that followed 1.2 million Norwegian births over two decades, women who themselves had been born prematurely had a higher risk of giving birth to premature babies. Taken together, the research suggests that people who were born early may want to make sure their doctors are aware of this fact when they themselves become pregnant.[60]

Repetitive pain, serious bacterial infection or maternal separation may cause changes in the neonatal brain that not only alter pain sensitivity but also increase anxiety, stress disorders, ADD and ADHD—conditions that too often lead to impaired social skills and ultimately to self-destructive behavior in later life. Based on two major studies, Dr. Anand theorizes that traumatic birth experiences imprinted on the brain account for rising rates of self-destructive behavior in adolescents, including suicide and drug abuse. He cites two major pieces of research. The first is an epidemiological study that correlated the birth and death records of male victims of violent suicides—death by hanging, gun, jumping or strangulation. Compared to matched controls who died of other causes, the study found that most of the suicide victims came into the world by forceps or vacuum deliveries and experienced resuscitation and other pain-inflicting interventions at birth.[61]

A second study correlated a cluster of coordinates—lack of care following birth, chronic maternal disease during pregnancy, and compromised breathing in the newborn—with increased risk of suicide in adolescents. This study made other interesting linkages as well. Sedatives given to mothers who didn't need them increased the risk for subsequent drug abuse in the offspring. The use of multiple doses of opiates, barbiturates and nitrous oxide during birth was linked to subsequent opiate or amphetamine addiction in adolescent children compared to those whose mothers had been given no such drugs. Anand and his colleague Dr. F. M. Scalzo say that there is a greater correlation between obstetric factors and later drug abuse in the offspring than between socioeconomic factors and addiction. Maternal smoking during pregnancy emerged as a greater risk for later criminal behavior in male offspring than demographic, perinatal and parental risk factors.[62]

Anand emphasizes that disruptive experiences during the neonatal period have a relatively greater impact than disruptive experiences later and set the stage for lifelong behavior patterns. This sensitivity has been documented in animal studies. In rat pups at six to nine days of age, sensitivity to pain is at its peak. This period in rat development corresponds to the development of the full-term human infant at birth. At this point, either too much or too little sensory stimulation can be toxic. Newborn rat pups (zero to seven days old) subjected to repetitive pain (which in

these studies was four daily needle pricks) showed a greater preference
for alcohol as adults, along with increased withdrawal and hypervigilant
behavior as adults. In a second group, pups were given injections that
caused inflammatory pain. This group showed decreased body weight
and decreased locomotor activity as adults. The females in particular
showed an increased preference for alcohol. Both groups of pups that
had suffered excessive pain in their first week of life had decreased pain
thresholds as adults. The researchers believe that pain experienced shortly
after birth probably altered both opioid receptors and the dopamine and
serotonin pathways that affect sensations of pleasure in the brain.[63] These
findings echo Felitti and Anda's ACE Study correlations in which the
bridge for many Kaiser patients between early adverse experiences and
adult disease was high-risk behavior in adolescence, including addiction
to alcohol and other drugs.

If animal studies are any indication, the perinatal period holds
tremendous potential for minimizing our vulnerability to trauma by simply employing birthing practices that honor the heightened sensitivity of
the immature nervous system. While the last few decades have seen
tremendous advances in obstetrical technology and minimizing infection
and related risks, there is still a great more we can do to minimize trauma
for babies and mothers during this crucial time without compromising
their physical safety or well-being.

CHAPTER 5

Little Traumas

Infancy and Toddlerhood

IN 2008 EVERY MAJOR tabloid publication ran multiple cover stories on Britney Spears. Her notoriety at that time came not from her undeniable talents as an entertainer but from her multiple arrests and embarrassing public confrontations that appeared to stem from either addiction or mental illness. When Spears was rehospitalized following yet another episode, *US Weekly* magazine published a report on the implications of this chaotic yearlong drama for her two little boys, then ages three years and sixteen months.[1] The overall tone of the eight-page article, entitled "Time Bomb," was dismissive of any long-range concerns for the babies who had spent the majority of their lives buffeted by the unpredictable winds of their mother's "condition"—still unspecified. Whatever was causing Britney Spears's problems—and there has been much speculation—the results for her boys included multiple separations from her, highly dramatic displays of emotion by both parents, involvement of the police and mental health systems, and frequent interruptions in their environment, not the least of which was the chronically unpredictable absence of their mother. The take in the media on the two boys' experience was not unusual. A New York child psychologist was quoted—in bold print under a photograph of Spears's youngest child being carried into Cedars-Sinai Hospital by an "unidentified adult"—as saying: "A trauma at this age is not necessarily going to stay with them."[2]

This is a widely held opinion about young children. Most people still believe—and understandably want to believe—that trauma experienced by infants and toddlers is erased over time, lost in the fog of early experience. The fact is that few of us remember our first or second birthdays, and most of us have only sketchy memories of ourselves at ages three, four and five. We may be able to conjure a few chimerical images of our old neighborhoods, houses, pets and neighborhood friends, or perhaps we know a family story about an early accident like a fall down the stairs or a finger shut in a door. But we have little conscious memory before kindergarten or first grade. Our first remembered stories that emerge from our own recall—as opposed to stories retold by older family members or imprinted from photographs—typically correspond perfectly with the time in our development when we have enough language to communicate a coherent story—about age five or six.

By this time our cortical brains are gaining some amazing new skills. Now coming online is our capacity to comprehend and to use words to communicate needs, wants, experiences and ideas, along with our growing ability to self-soothe and to control the primitive, impulsive behaviors of our baby selves. We have some basic interest in and even ideas about why people do things: most of us as kindergarteners begin to observe motivations, our own and other people's, at least on a simple level. We may have remarkable recall of some events from the past, though they are mostly remembered through our relatively new understanding, including the fact that we were "just a baby" when certain things happened.

The majority, however, of our remembered images from the past, once so vivid to our less developed selves, lie buried without words, stored within our cells as sensations or feelings in preconscious memory. Although we do in fact store memories in several unconscious memory systems, and even though they continue to exert an influence on our feelings and behavior, only the most recent of these memories are available to us in the new format of words as we enter kindergarten. Since few of us will consciously remember our first two years of life, it is not surprising that we share an amnesiac view of infancy and toddlerhood.

This was certainly the mind-set of a group of defense attorneys in the 1990s whose job was to defend an infamous serial killer. According to the defendant's brother—a highly respected professional who has devoted his life to defeating the death penalty—when the defense team interviewed their mother, she told them some puzzling things about the defendant as a baby that they subsequently passed over because they seemed irrelevant. Now ninety-five years old, this mother, Wanda Kaczynski, remains as passionately adamant today as she was all those years ago that her baby was transformed by trauma.[3]

She says that at birth and for his first months Ted was a normal if somewhat fussy baby. He was her first, and both she and Ted's father were delighted with him. But at nine months Ted was hospitalized with a terrible case of hives. In those days, during the last world war, hospitalized babies were routinely separated from their family; regardless of their protests and their grief, parents were preemptively shut out. Wanda says that for the week that Ted was in the hospital, not only was he isolated from other children, but she was only allowed to see him for one sixty-minute visit during the course of his stay. When Wanda arrived to finally bring Ted home, she says she found herself holding a different child. Baby Ted was now limp, "a rag doll." No longer did he fuss for attention. He wouldn't look at her and was frighteningly quiet. The baby was, of course, Ted Kaczynski—who two decades later would be known as "the Unabomber."

It would be hard to attribute serial murder and the kind of hatred that Kaczynski voiced in his "Manifesto" to early trauma alone. Nevertheless, it is hard to understand what factors could have so warped the brain of this child who grew up in the same family that also produced as compassionate and sensitive a soul as David, who brokenheartedly led authorities to his brother. This was a family in which the parents, first-generation children of Polish immigrants, worked hard to gain educational opportunities for both of their sons and, according to David, had no greater priority than being "good parents."

Ted's baby book provides clear evidence that Wanda Kaczynski was aware of the drastic changes in her baby and, on some level, knew that all was not well. She explored several child development resources as Ted was growing up, but discovered little that could adequately explain his behavior. When Ted was a kindergartener, she looked into Bruno Bettelheim's work with autistic children at the University of Chicago. After observing Bettelheim's laboratory classroom, Wanda decided that

Bettelheim's manner was too abrupt and cold, and she was unwilling to put Ted into such an environment. She also concluded that her son was not exactly like the children Bettelheim worked with. Unlike those children, Ted, though he had been slow to speak and never went through the typical baby babble or reciprocal exchange of cooing and syllables, spoke in complete sentences when he did begin to talk. This pattern also manifested in his large motor skills. Wanda says that "Teddy" never crawled or scooted but went straight to walking sometime during his second year. He grew into a child who didn't relate well to other children and was uncomfortable with anyone outside his family.

By late grade school, Ted actively disliked other children and interpreted their behavior as malevolent. He objectified other people and had no need for connections outside of his family circle—a trait that grew gradually into abject hatred of people in general. Before his incarceration, Ted told his brother, David, that he was the only person he ever really loved—although there is a poignant little Mother's Day card folded into Ted's baby book, consisting of a child's crayon drawing of a boy and several drawings of "mother" with the words "I love you, I really love you." Ted has no contact now with anyone, not even his family, with whom he severed all communication shortly after his sentencing.

David Kaczynski continues to devote his life to eliminating the death penalty, a fate his brother narrowly escaped in spite of the fact that authorities had promised David that they would not pursue this course if David led them to his brother. His father is now deceased, and his mother is frail, though her mind remains sharp and clear. Among David's ongoing efforts to right some of his brother's wrongs are an abiding friendship with one of his brother's surviving victims and a commitment to facilitate the telling of his mother's story.

Wanda Kaczynski maintains that Ted's hospitalization as a baby made him a different child. And regardless of other factors that also may have contributed, it is evident from her description of Ted's affect and behavior at the time of his discharge that he had indeed been severely traumatized and never received treatment. From a baby's perspective, the absence of healing and repair is, in itself, a source of retraumatization. In the absence of knowing that such events constitute a tsunami in a baby's brain, which in turn triggers a tidal wave within the child's body, parents may witness ongoing reverberations of trauma throughout the child's development. Unrecognized and unhealed, trauma may generate a horrific breach of trust in the parent that may extend to the rest of the human community. Stored without words and frequently viewed as "best forgotten" by well-meaning parents who have no idea how to bridge the gulf, this experience is often the beginning of a serious deviation from a normal developmental path for a child. Little traumas could accumulate and interact with genetics to create a path of emotional and physical disease.

The good news is that we now know much more about this trajectory than we have known historically. We are discovering what constitutes trauma from a baby's perspective, what behavior might signal serious difficulty, even how to measure whether such trauma is occurring, and, most important, how to begin to reverse such problems. We may be seeing the dawning of an age when the kind of ignorance that shaped Ted Kaczynski's life could become a thing of the past. Although Kaczynski's story is an extreme one and may have been complicated by an undetected developmental challenge like Asperger's syndrome (a diagnosis not yet available in Ted's childhood), the newest science examining trauma and its impact on development, especially in the presence of genetic challenges, could greatly alter the trajectories for such children through early detection and intervention.

Having interviewed hundreds of clients during the thirty-three years of my practice, I know that people often begin talking about their histories, so many of which are riddled with ill health, by reassuring me that they had wonderful parents, an intact family, and certainly no abuse. And they are right. Although child abuse and neglect are obvious forms of trauma, we are learning that it doesn't take either to constitute trauma for a very young human. For a baby or young child, emotional trauma may not look like the catastrophes we typically connect with the word *trauma*. In fact, most trauma in early childhood—from the First World to the Third World—is accidental, and its major cause is ignorance. Few people set out to traumatize their babies, and most of us want to protect them. But our culture is still largely blind to the sensitivity of the human nervous system, particularly as it is being formatively built.

In this chapter, we look at the kinds of experiences that trigger trauma for babies and toddlers, between birth and their third birthday. And we examine some common experiences, ingrained in our culture, that can have a traumatic impact on developing nervous systems. Our discussion of these primarily accidental and well-intended experiences is not meant to cast blame. The resulting trauma may be a side effect of an essential life-saving practice, like surgery, or of the isolation of a baby whose malady may be contagious, like the case of hives in baby Ted. Often these experiences are tied to economic necessity; many single parents, for instance, have to work and can only afford low-quality child care. Nevertheless, efforts to identify sources of trauma in very young children and recognize its effects need to be integrated into medical, social, educational and mental health policies, practices, and protocols.

Until very recently, few have understood what constitutes trauma from the baby's perspective, including many well-meaning and well-educated professionals—physicians, judges, lawyers, businesspeople, teachers. It was certainly trauma when Wanda Kaczynski's baby boy, a nursing infant of nine months, was suddenly removed from her care, was placed with strangers in a hospital, and had his hands tied down to a gurney to keep him from scratching. Baby Ted was no doubt fed and changed. But he did not see his mother except for one hour that week.

His treatment in what must have been one of the best hospitals of the time is ironically reminiscent of the sterile protocols rigidly followed by the Romanian orphanages that were so extensively covered by the American media in the 1990s following the fall of the dictator Nicolae Ceaușescu in 1989. Americans were shocked as the media ran photos of the Dickensian conditions in these orphanages that were overflowing with children who had birth defects, developmental delays, or families too poor to support them. Appearing on television and in magazines and newspapers across the world, the babies were photographed listlessly sitting or lying in barren metal cribs, often four to six in a crib, apathetically staring into the camera. Their little faces, graphically documented in Nobel Prize–winning *Time/Life* photographs, were very unbabylike: drawn, eyes dull, often crossed and heavy-lidded, expressions devoid of smiles or pleasure of any kind. The babies were apparently clean and fed but lacked any form of loving human emotional interaction. When they cried, no one came to comfort, rock, sing and soothe. There was no holding, singing, reading, teaching, playing or laughing—no reciprocal emotional exchanges of any kind. As adoptees in the United States and Canada, the Romanian orphans caught the attention of several researchers when they showed limited or nonexistent capacities to attach and interact with their adoptive families or were slow to talk and learn. Their prognosis varied with their circumstances: their age when they were left in the orphanages, how long they stayed, their temperament, their intelligence, and the presence or absence of developmental challenges. It was certainly trauma to these babies to be denied social and emotional interaction. But Romania wasn't alone in its ignorance about what constitutes trauma for baby humans.

The biological changes that characterize trauma can be very difficult to discern in the behavior of the youngest children—mostly because we know so little about normal emotional behavior in earliest life and the kinds of behavior that may indicate serious distress. Recall Annie Rogers's definition of trauma as "any experience which by its nature is in excess of what we can manage or bear."[4] We do know that it is particularly easy to overwhelm the emotional capacity of a baby whose nervous system is new, unformed and raw—without learned defenses of any kind.

It is this issue of the *emotional* vulnerability of the immature system that is so commonly overlooked in our culture. We seem to understand, to the point of queasiness, the *physical* vulnerability of the newborn. Many new dads and people inexperienced with babies are so stricken by the fragile appearance of the new baby that they are afraid to hold this tiny bundle of unpredictable and unregulated visceral responses—afraid they might drop the baby or squeeze too hard or become the target of unanticipated bodily fluids. But as concerned as we may be about the newborn's physical fragility, we are often inured to the newborn's emotional vulnerability, assuming that our newest family member is unaware and therefore insulated against harsh words and arguments, expressions of anger, frustration, grief, fear or tension from adults around the baby.

In fact, the opposite is true. Babies are wired to be exquisitely tuned to both positive and negative emotions, which they literally absorb from the adults around them. Their limbic brains are set to be particularly receptive to the rhythms and moods and tones and feelings of the world around them, especially those of their main caregiver. This relationship provides the blueprint for the baby's own emotional regulation and future expectations of relationships, especially intimate relationships with other people. Built-in sensitivity and malleability to the environment is a great asset for our survival, but it presents an immeasurable responsibility for adults, whose commitment to nurturing the young is lengthy and labor-intensive.

Earliest development is not only a period of singular vulnerability, it is also our best opportunity to build ballast against later trauma. Before there is a sense of self, before a buffer has been built from comforting experiences that enable a little brain to know that emotional balance *can* be restored after a frightening or painful event, children are vulnerable to having their nervous systems dysregulated. Early positive experience is protective. But, as Phyllis Stien and Joshua Kendall point out, because immune and endocrine functions are so closely tied to brain functions through the HPA axis, "tragically, one single traumatic experience is enough to alter brain functioning."[5]

BRAIN BASICS

Brain development basically occurs from the bottom up. The first part of the brain to develop is the most primitive: the *brain stem* controls the most basic functions of the body over which we have no conscious control, like blood pressure, body temperature and respiration. Next comes the *midbrain*, which controls bodily functions over which we have some awareness and control, like appetite and sleep. Then the *limbic brain* develops: this is the seat of emotion and impulse. Finally, developing the most slowly, is the *cortex*, the outside layer of the brain that is the seat of logic, planning and rational thought, the "executive functions." It is worth noting that complex cortical skills, such as self-control and connecting what we do today with what may happen a week from Friday, are still very much coming online throughout adolescence—which is why so many teenagers have such difficulty with these capacities.

Use and stimulation are brain food, essential to fulfilling brain potential; development reflects the quality and quantity of these experiences. The younger the child, the more the primitive areas dominate. As the higher areas of the brain develop, the cortical brain gradually begins to modulate and control the more primitive lower brain. What this means for a child is that any experience during critical brain development that *reduces* the developing brain's cortical capacities (like chronic emotional neglect after birth or exposure to alcohol in the womb) or any experience that *increases* the responsiveness of the lower brain (such as exposure to chronic fear) can help set the stage for ill health later in a child's life.

We are living through a time of great irony (though perhaps this has always been true): on the one hand, we have the technology to create graphic images of the human brain and how it works, and we recognize that behavior—all behavior—is rooted in the structure and chemistry of this physical organ. We also know that, especially in earliest development, the brain is designed to literally shape itself in response to the environment. But accompanying these high-tech discoveries about our species has been such a rapid rate of cultural change that families have not been able to keep up, let alone replace, the relatively low-tech supports essential for quality child-rearing, particularly the time spent in one-on-one, face-to-face contact. We continue to pay a higher and higher price for this oversight. Like Humpty Dumpty, all of the systems designed to remediate the fallout of early developmental trauma—from early intervention and special education to addiction treatment to mental health systems to prisons to overwhelmed health systems—are together hard-pressed to repair the consequences and put the child together again.

OF JOEYS AND BABIES

Under the best of circumstances, with basic needs for food, shelter and warmth already assured, we face two major challenges at birth and in the first weeks of life. The first is to regulate arousal so that we are not constantly overwhelmed, an essential prerequisite to all that follows. The second is to adequately focus on the psychedelic world of never-before-seen colors, never-before-felt touch, never-before-discerned sounds and tastes and movements that greet us in the world outside the womb. If we are too sleepy, we do not get the brain food we need. If we are too excited, we cannot absorb and learn from what we see, hear, touch, taste and feel. And we can't balance any of these experiences by ourselves.

As dependent as we are on an adult for food and warmth and physical protection, we also depend on an already organized and mature nervous system to guide our yet unfinished one. It takes a modulated adult to monitor and balance newborn over- or under-arousal and help regulate our raw and reactive systems so we can maintain some degree of homeostasis. If we are overexcited, we need swaddling, soothing and reduction of stimulation. If we are drowsy, we may need to have our feet tickled, or to be picked up over a shoulder, jiggled, and talked to in a lively voice. The lack or the failure of a competent attachment figure is certainly trauma for the newborn, who is left at the mercy of a nervous system that is far too easily dysregulated. A committed adult is crucial to teaching each of us how to drive our brand-new system, and in fact, the skills of that adult are all that stand between us and circumstances that can easily take us off track physically and emotionally.

This is the price we pay as a species for the unique adaptability of the human brain. Like baby kangaroos, humans are born strangely unfinished. And like little kangaroos, baby humans benefit from a pouch-like environment, a homogenized and sensitive extension of the womb, at least for a few weeks, while learning the basics of physical regulation.

Newborn kangaroos begin life just under one inch long, hairless, pink, looking much like little grubs or embryos. Birth for this creature is a short three-minute crawl from the birth canal up over the mother's belly, a journey that snaps the trailing umbilical cord and releases the baby to tumble into its home for the next few months—the mother's pouch, where it latches on to a teat. The baby, now called a "pouch embryo," continues to develop as if it were inside the womb. This transfer of environments is so subtle that until relatively recently biologists studying marsupials thought that kangaroos began their lives in the mother's pouch. The embryonic-looking newborn appears to grow as one unit with the teat, which swells upon initial contact so that the baby and the teat are permanently attached for months in the pouch. It is then ten months or so before the "joey" emerges for the first time for a short peek at the larger world. It goes through another eight months of gradually extended exposure while acclimating to life outside the pouch. During this time, the baby still sticks very close to its mother's side.

Baby humans obviously have a very different arrangement. Nature made sure that the rental agreement for the ideal real estate in the human womb expires relatively suddenly after forty weeks. This perfunctory expulsion ensures that the human baby's large head will not outgrow the birth canal. But like the newborn kangaroo, humans aren't nearly ready at birth for life in the "outback." Still dependent on a mature adult for the soothing that leads to regulation of breathing and heart rate, sleep and feeding, warmth and mobility, the newborn baby needs protection not only from outside predators but also from the stress of being overwhelmed internally by unmodulated stimulation entering the immature nervous system through untuned senses.

Between the womb and life in the big world, baby humans have a similar need for a transitional environment that ensconces mother and baby in the still-unfinished process of readying an immature nervous system for life in the human community. Ideally the attachment relationship between the baby and the mother (or another consistent caregiver) acts as the baby human's "pouch." This relationship provides the envelope of ongoing protection and gradual titration of stimulation that insulates against physical and emotional trauma to the newborn brain. The loss or disruption of this protection is the major threat of trauma to the baby after birth.

EARLY SEPARATION FROM THE PRIMARY ATTACHMENT FIGURE

The first glimpse of trauma resulting from the effects of maternal separation were captured for the world in the photographs of Harry Harlow and his little monkeys in the 1950s. Following World War II, after John Bowlby and René Spitz published their work on the devastating effects of institutionalization on very young children, Harlow began a series of experiments with baby monkeys separated from their mothers. He had little monkeys choose between cloth-covered and wire surrogate mothers. Only the wire "mothers" were equipped with food, yet all the babies preferred the cloth-covered "mothers" to cling to when threatened, demonstrating that mothering meant more than nutrition to little monkeys. Also, regardless of the surrogate, all of the monkey babies separated from their biological mothers had a harder time with learning and socialization than little monkeys left with their mothers.

What is less commonly known is that, unlike the little monkeys who were later assigned the cloth surrogates, monkeys forced to live exclusively with wire surrogates also showed many signs of physical distress. They had difficulty digesting milk and developed chronic diarrhea. Although Harlow's findings would prove controversial, he was a pioneer in the emerging science of attachment and its impact on basic physical functions. Today hospital protocols across the nation honor the importance for infants of normal sensual contact with their mothers. Maternity practices like skin-to-skin placement of the newborn on the mother's chest and the phasing out of newborn nurseries for healthy newborns (mothers are now encouraged to have their newborns "room in" after birth) reflect a revolution in our understanding of the importance of early attachment.

Since the 1950s, baby rats have also provided us with a clear picture of the effects of maternal care and its loss. Based originally on the work of Dr. Seymour Levine, Dr. Michael Meaney's research at McGill University in Canada has found that high-licking-and-grooming mother rats produce pups that are larger, less stressed and healthier as adults than the pups of mothers who are less attentive. By the time Meaney had observed this outcome in rats, earlier research had already shown that when newborn rat pups are separated from their mothers they become highly sensitive to stress, less able to explore novel situations, and more vulnerable to stress-related diseases. From immune disturbances to cardiovascular disease to depression and neurodevelopmental disorders like Alzheimer's, maternal separation of newborn rats has repeatedly been shown to greatly increase the likelihood of disease.

But Meaney's research provided a new twist. He left some of the newborn rats with their mothers, separated others from their mothers for fifteen minutes a day, and took a third group away from their mothers for three hours a day. When he looked at the babies as adults, he found some unexpected differences. Pups that had been separated for only fifteen minutes a day actually did better than those that were left with the mother. The separation was brief, yet long enough that their anxious mothers gave their pups extra attention when reunited. Extra attention by the mother—licking of the anal-genital region—enhances rat pup growth and decreases the release of cortisol. Meaney discovered that the level of care a mother rat gives her babies actually changes the DNA in certain genes involved in the stress response. Highly nurturing mothering switches on a gene that controls the production of cortisol. This does not happen with low-licking mothers, whose offspring produce higher levels of cortisol when stressed.

It appears that the mothering style in rats is passed down from one generation to the next, not by genes, but by experience: the offspring of low-licking-and-grooming mothers that are raised by high-licking foster mothers will emulate the foster mothers' high levels of grooming with their own pups. These positive results from fostering are time-sensitive: the pups must be moved to a high-licking foster mother very early and left without further disruption.[6] Recent research on mice validates the relationship between maternal separation and disease. Pups routinely separated from their mothers at one to fourteen days after their birth are far more vulnerable to viral lung infections than pups left with their mothers. Researchers attribute this difference to the dysregulation of cortisol.[7]

Alicia Lieberman, chair of the Department of Psychiatry at the University of California, San Francisco (UCSF), specializes in a relationship-based approach to treating mothers and the youngest victims of domestic violence: their infants, toddlers and preschoolers. Lieberman defines trauma as anything that results in the disruption of the "secure base"—essentially any experience that seriously disrupts the physical and emotional balance and security provided by the child's primary relationship with an adult, typically the mother.[8] It is stunning to imagine the extent of trauma for infants in our nation if Lieberman's definition of trauma is accurate. The dearth of quality child care alone poses a great threat to the well-being of American children, to say nothing of the issues that Lieberman addresses in her work, namely, the impact of domestic violence, child abuse and neglect. In a 2008 interview following a presentation before the Dalai Lama in Seattle, Lieberman said very simply: "One of the most important findings over the last thirty years is that there is no such thing as a baby not noticing what is happening."[9]

Unfortunately, there are still many routine practices in our culture that traumatize children by inadvertently ignoring this fact. We have already touched on medical practices. But that is only the beginning. There are legal practices, like custodial and visitation protocols, and social service practices that dismiss the reality of trauma in the training and preparation of parents who foster and adopt seriously affected children. Foster children rarely receive adequate trauma therapy or other means of healing trauma, including access to music, athletics and dance. There are also many protocols that lead to maternal separation, such as welfare programs for impoverished families that punish women who try to stay home with their babies, particularly teen parents who must return immediately to work or pay the price of losing benefits. Then there is the uneven quality of child care in our nation, a particularly worrisome issue when it comes to care for infants. When the full implications of this information on the impact of early trauma sink in and we comprehend what constitutes trauma from the baby's perspective, we begin to look differently at many taken-for-granted practices.

ADOPTION

Although adoption is an invaluable solution for children and families, it is nonetheless likely to be traumatic for a young child, even when it occurs just after birth. In the intensity of the exchange from one mother to another, we may underestimate the sentience of the preborn and the subtle factors at play that may have already affected the baby. In the womb the fetus became familiar with the biological mother's voice, tasted her fluids, felt her rhythms. The biological mother is likely to be stressed by the decision to release her child and by the circumstances in her life that led to her decision. Grief and apprehension rather than joy are often likely to accompany the birth of a baby destined for adoption, and these feelings may contribute to birth complications. Sometimes, because the biological mother has hidden her pregnancy, the fetus is undernourished or has been exposed to alcohol, nicotine or other deleterious substances. Mother may not have received good prenatal care, and she may be depressed. On top of all these factors, birth itself may be traumatic, complicated by breech, cesarean delivery, induction or the use of suction or drugs.

Even when birth has gone well and capable adoptive parents are standing by, trauma for the baby can be unavoidable. Because this is such a joyful time for adoptive parents, the issue of emotional trauma to the baby is rarely considered. The child may experience several "little traumas" in the course of the adoption, most of which can be repaired by stable, loving care. But for many adoptees and their families, the hidden consequences of prenatal and perinatal trauma surface in later behavioral problems, when they are often ascribed to genetics. Early trauma often surfaces in grade school as attentional and focusing problems or anxiety or aggression, or it may emerge in preadolescence or adolescence as addiction, sexual acting out or self-destructive behavior. Adequate preparation of adoptive and foster families through education about trauma and its impact on the brain can make a huge difference. For young children who have undergone major breaks in attachment, child care arrangements are particularly important; exposure to multiple caregivers risks additional trauma. For an adopted or fostered infant, as for any infant, secure attachment—stable, committed care by someone who loves the child unconditionally—is the best therapy.

In the United States, children adopted or fostered from intensely traumatic beginnings are often diagnosed with reactive attachment disorder (RAD), which requires a strong comprehension of its roots in trauma to heal. Both family and individual therapy for the child are likely to be needed to prevent more intense problems down the line. Education on the effects of trauma is essential for legislators, child welfare professionals, judges, therapists and adoptive and foster parents. Therapeutic stipends or trust accounts should be the rule rather than the exception for adopted and foster children placed through public and private agencies, regardless of their age at adoption; this is especially important for children recognized as victims of any of the adverse childhood experiences delineated in the Kaiser ACE Study.

FOSTER CARE

Separation from biological parents—even parents who have been abusive or neglectful—typically adds more trauma to a series of traumas already experienced by the child. Though attachment may have been fraught with pain, these are children separated from their primary attachment figure and the same conditions apply: loving homes secured through sensitive screening and preparation provide the best possible therapeutic opportunity. Foster children generally require the "superduper" care described by Charles Nelson as necessary to meet the needs of neglected Romanian orphans. Unfortunately, love alone is not enough: these families need exceptional skills and equivalent patience. The children, especially those from child protective services, suffer high rates of PTSD. When therapy is inadequate or nonexistent, especially after a child has had multiple placements, many of these children enter adulthood with trauma symptoms in full force as they begin their own families. All of the same needs for support and training apply to foster parents that apply to adoptive parents: this is a great opportunity to break the trauma cycle for many children.

DIVORCE

There are two diametrically opposing views about the impact of divorce on children. One argues that parents should stay together "for the good of the children." The other argues that parents who are unhappy with each other should divorce "for the good of the children." Experts are divided. Dr. Judith Wallerstein, coauthor of *The Unexpected Legacy of Divorce: A 25-Year Landmark Study*, concludes that adults who were children of divorce struggle with anxiety and loneliness and problems with love and commitment.[10] Her interviews led her to believe that living in a broken family leaves kids poorly prepared to form intimate relationships and that childhood insecurities come home to roost in subsequent relationships. Another expert, Dr. E. Mavis Hetherington, formerly of the University of Virginia, has written a very different book on the topic: *For Better or Worse: Divorce Reconsidered.*[11] Hetherington found that while it is true that 20 to 25 percent of children whose parents divorce are at risk of lifelong emotional or behavioral problems (compared with only 10 percent of those whose parents stay married), these figures still mean that 75 to 80 percent are functioning well or at least in the normal range. Both Wallerstein and Hetherington agree that, as adults, children whose parents divorced are much more likely than children of nondivorced parents to become divorced themselves; data show that women are more than twice as likely to divorce, and men 30 percent more likely. In the 1960s and 1970s, as many people sought individual fulfillment and felt they were providing their children with models of individual freedom, divorce became the norm. But this trend has been ebbing. At the divorce rate's highest point, 50 percent of first marriages and 60 percent of second marriages reputely ended in divorce in the United States, but those figures are dropping. Still, between one-third and one-half of the children in this country have weathered a divorce.

So what does divorce mean from the perspective of a very young child? The term "divorce" connotes a specific legal event, but anyone who has experienced this process knows that in fact divorce occurs in stages, some of which happen long before any legalities are involved. Assuming there is no physical violence, children in most families facing divorce feel that something is awry; before parents separate, the children may sense increasing tension and observe subtle or not so subtle differences in parental behaviors—one parent absent for long periods, tears, hostility, irritability, tight voices, less laughter or time for play. Often this first phase goes on for quite a while before it culminates in the decision to separate or divorce. From start to finish, divorce is generally a painful process for everyone and has recognizable stages: pre-separation conflict, telling the children (if they are old enough to understand), moving out, new neighborhood and new routines, and the introduction in many cases of new adults or stepparents.

Potential for trauma—and the deterioration of a child's ability to cope—lurks at each of these crossroads. Remember that, for a baby or a child under three, these experiences are recorded as somatic rather than verbal or rational memories. Many variables influence how traumatic these events are; especially critical factors include the level of conflict between the parents, the vulnerability of the child owing to prior history and temperament, and the degree of secure attachment to one or both parents. Mother's and father's individual adjustments are also strong influences, especially their ability to rise above their own negative emotions and prioritize the interests of the children. Also key are the compromises and agreements they negotiate concerning a parenting plan and discipline, their commitment to communicating civilly and openly with each other, and the presence or absence of additional stressors, such as economic hardship.

Does all of this register with a child? Absolutely. Is the process of divorce always traumatic? Not necessarily. Is it the same for all little ones? No. Adjustment differences also depend on the child, his or her temperament, what has gone before, and who is with the child as he or she goes through these experiences.

Infants and toddlers may not understand conflict, but they feel the tension. And even tiny bodies—especially tiny bodies—react to extreme negative emotions and drastic alterations in parents' moods and behavior. The baby may show a loss of appetite or an upset stomach, increased spitting up, gas or colic, fussiness, changes in sleep and elimination patterns, or difficulty being consoled or soothed, particularly if a parent is tense. Toddlers may regress, especially if one parent moves out or is gone from a familiar routine like bedtime story or playtime after dinner. A potty-trained three-year-old may begin soiling her pants, a weaned two-year-old may demand the breast or bottle, or a child who has been sleeping in his own crib may refuse to stay there or wake several times during the night.

Until about six months, infants don't understand that people they can't see still exist. If a baby is attached to one parent as a primary caregiver and that parent is lost to him or her, the baby may respond much as if the parent had died. She or he may grow sad and withdrawn, and all vital functions, sleeping, eating and playing will be affected. If the parent is unavailable to the baby for an extended period and then returns, the baby may behave as if the parent were a stranger. For youngest babies, it is hard to remember and form close relationships with people whom they don't see often. Between six and nine months, when it is normal for babies to develop stranger anxiety, a child may respond to an infrequently visiting parent with lack of recognition, anxiety and fussiness. Frequent contact with both parents is important, though a fifty-fifty split is a poor choice for a nursing baby or toddler.

Sometimes toddlers or preschoolers manifest their grief by becoming clingy or angry and aggressive, biting, or losing formerly learned social graces like listening, taking turns and sharing. There may be an increase in accidents and nightmares. Wanting to stay close to the parent, they may refuse to attend child care or preschool. Tummy aches are common among children trying to regain parental attention, as are regressions in getting dressed and other self-care habits. The children of separating parents need constant reminders of adult responsibility for decisions and reassurance of ongoing love and access to both parents. Parents are, of course, overwhelmingly stressed themselves at the same time and need the support of friends and family and, whenever possible, professional support as well.

Children undergoing upheaval need routines that are familiar and comforting, a sense that the rhythms and patterns of their world remain predictable. This is not a time for changes that can be postponed. Sticking to the same people—child care providers, schools, neighbors, doctors—is helpful, and even toys and clothes should be changed as little as possible. Most of all, parents can provide the scaffolding that minimizes trauma by giving the child time for extra contact with both parents. Grief for a child is normal, but trauma is not. Trauma can be minimized, however, if divorce is handled sensitively, mindfully, and gradually (see the suggestions listed in appendix A). Even among the youngest children, the generalized reaction to grief is sadness, but in trauma the generalized reaction is terror. Grief is easier to observe in children's behavior, while trauma is often obfuscated—expressed as withdrawal, regression, anger, even self-abuse. Grief is relieved by talking—trauma is not. In fact, talking may be difficult or impossible for young victims. Trauma is slower to dissipate than grief and is best dealt with through movement that allows the discharge of "frozen" fight-or-flight energy. A grief-stricken preschooler may be angry but is generally not violent to himself or others. A traumatized child may be abusive to another child or to himself (for example, the preschooler who gets into frequent "accidents"). Although both grieving and traumatized children may have nightmares, the grieving child's dreams are generally about the lost parent, while the traumatized child's dreams involve being victimized, often horrifyingly so. If a child is hypervigilant, numb, dissociative or experiencing flashbacks, the child is in trauma.

Trauma to children during divorce or separation is most likely to occur when their parents are too caught up in their own emotions to adequately protect the children. Intense anger, feelings of rage, rejection, fear, frustration and revenge can consume parents' attention. Children, even babies, experience trauma when they hear or see verbal or physical or emotional violence, when the separation occurs suddenly without preparation, or when a close bond is suddenly lost or threatened. Even in the absence of a single overwhelming event, trauma may result when, as a consequence of the decision, the child is plunged into less nurturing or unstable child care, when parents are in chronic emotional distress, or when the child's access to a protective relationship is diminished. This is particularly true for a child who has been previously traumatized or who has a sensitive temperament.[12]

The trauma of divorce is greatly exacerbated by the gap between what we know about children's needs and how the legal system persists in prioritizing the rights of parents over the rights of children. The King Solomon story plays out in American courtrooms every day as parents request of judges that their children be "cut in half" to honor their "equal rights" to custodial time. For nursing infants and toddlers, for whom a secure base with one parent is paramount for the first eighteen to twenty-four months, this process alone is trauma. Children deserve access to both parents, and vice versa. But under optimal circumstances, a baby is breast-fed and remains with the mother long enough to establish that crucial physical and emotional foundation. This is not a time for mothers to monopolize the attachment process: ideally fathers have hours with their babies on a daily basis. But if agreements can't be negotiated peacefully, there may have to be some sacrifice of the father's "rights" until the baby is old enough to spend time in two households. Loving fathers and informed judges, like the biological mother in King Solomon's court, will not allow a baby to pay the price of ensuring "equal rights" for the adults. A parenting plan created with an experienced therapist can often prevent the ongoing emotional and legal battles that compound trauma.

DEATH OF A PARENT

The death of a parent is a major source of trauma for babies and children. The degree of trauma a young child experiences is commensurate with the parent's closeness to the child, as well as the availability of another consistent and loving adult who can form and maintain a close relationship with the child. When a second parent is still available to the child, or adoption by another adult is a possibility, or when attachment is nurtured and regained with a surrogate parent, trauma may be mitigated. But for many children of any age, the trauma of parental loss or abandonment lingers unrecognized until it appears as a behavioral problem such as addiction or as depression, or when the child is diagnosed, often mistakenly, as having ADD, ADHD, or a learning disorder.

Infants who lose a parent commonly reflect both anxiety and depression, which often go unrecognized and may in turn surface as allergies or various anxiety-related diseases, including asthma, eczema, digestive problems and headache. The death of an involved parent is always traumatic, even when the resources for supporting the child are firmly in place. This is especially true if the parent's death leads to a loss of status, exposure to poverty, or shame. In the face of this reality, it is overwhelming to consider the number of children across the world affected by this form of trauma. More than 163 million children are orphans from causes ranging from HIV/AIDS, poverty and family breakdown to war, natural and man-made disasters and armed conflict.[13] Many of these children are living not in orphanages or even under a roof but on the street, abandoned by family and community. In Africa, according to UNICEF data, 12 percent of all children were orphans in 2010, as were 6 percent of Asian children and 5 percent of children in Latin America.[14] If all of the world's orphans today joined hands, they would span the globe. The growing number of children orphaned across the world carries frightening implications as the trauma of one generation echoes the last and adds to that of the next.

CHILD CARE

Few topics raise parents' blood pressure more than the ongoing rancor about the placement of young children in nonparental child care, which, in spite of our ambivalence, has become the norm rather than the exception. Care of children by someone other than parents takes place in many different settings, ranging from care by grandparents to care by a licensed home provider to center-based day care. Each form of care can be advantageous to a child, depending on the fit between the child's needs and the skills of the provider. Each form of care can also be devastating and result in chronic stress and even full-blown trauma. Certainly nonparental child care need not categorically cause fear or trauma for children. Studies using cortisol to measure infant and toddler stress are a window to understanding how to prevent trauma to children in day care.

Cortisol is Dr. Megan Gunnar's favorite hormone. A noted University of Minnesota researcher on stress and child care, Gunnar has created what she calls "the tasting game." She gives rolls of cotton dipped in flavored sugar crystals to toddlers to suck on, stimulating the secretion of saliva. Removing the saturated cotton rolls from the children's mouths, the cotton is whirled in centrifuges to determine the level of cortisol in each child's saliva. The results have revealed some clear realities concerning stress and child care.

In babies and toddlers who are cared for at home, cortisol levels are highest in the mornings and become progressively lower as the day wears on. But in children of the same age placed in day care, cortisol levels too often increase as the day goes on. The good—and perhaps most significant—news is that a secure attachment relationship with the mother or a primary figure can normalize a baby's cortisol levels. Beginning at four to six months of age and extending through the second year of life, strong maternal nurturing behaviors (similar to those high-licking-and-grooming mother rats) mitigate stress for babies in alternative care. For this to happen, however, it is critical that the mother or main caregiver develop such a relationship with the baby prior to separation.

Gunnar's research confirms that a securely attached baby who has been well attended by a sensitive and responsive caregiver may continue to cry and exhibit distress at separation from the mother, but by around six months of age will have about the same baseline cortisol levels in the alternative care setting as those taken at home. For insecurely attached children, by contrast, even minor emotional challenges raise cortisol levels. Gunnar found that 70 to 80 percent of toddlers and preschoolers in day care and preschool showed increasing levels of cortisol throughout the day during full days spent in day care, suggesting a loss of adequate HPA axis regulation.[15] This was especially true for children who were shy or fearful and for those who had difficulty regulating negative emotions. As in rat pups, the security of early nurture appears to bestow a buffering effect, a moderation of children's HPA axis, provided that care is of high quality. Factors that increase cortisol levels are the ones we might predict: poor quality of care, excessive hours in care, multiple arrangements (being placed in more than one setting or having a history of several care arrangements), or being placed in a center at too young an age or with a difficult history or reactive temperament. For slightly older children, a lack of peer acceptance also increases cortisol levels.

Drawing on cortisol measures and using other tools as well, researchers have conducted several large studies of child care that have been hard for them to publish and whose findings have been difficult for families to accept and digest. The result is that out-of-home child care, a crucial support for families—and one that is only increasing in need and use—has not yet caught up to what we know about what children need for optimal development. An industry that could and should be the foundation of our nation's educational system still lacks the standards and public support it deserves, and the education, training and salaries of child care workers remain subpar. As a result, many children are being traumatized and retraumatized by the gap between what we know and what we do.

Research points urgently to the importance of time and timing. Simply put, the younger the child, the greater the risk that alternative care, particularly center care, will have harmful effects. The Australian journalist and social philosopher Anne Manne discusses this issue in her book *Motherhood*.[16] She speaks of our political refusal to honor the research on early development and our rationalization of inferior but convenient options for child care as "the McDonaldization of childhood." A strong advocate of the Scandinavian policy of paid, job-protected maternity leave for both parents and paid sick days per child of up to 120 days per year, Manne is an eloquent voice for governmental backing of the option for parents to stay at home with their own children if they so choose. She also distinguishes between "nonparental" and "nonfamilial" care.

At the heart of this research is nonfamilial care—care provided by someone unrelated to the child who is paid (usually at a low wage) and may not be emotionally attached to the child. The many grandmothers and aunts and grandfathers and other extended family members who also look after children outside the child's home typically do not factor into this type of care. Some of the studies analyzing the care of young children by paid caregivers who are unrelated to the child look only at center day care as opposed to homes. This is one of several variables to keep in mind as we consider the results of this research. It is important to also remember that alternative care is the norm, not the exception, for young children in today's society; this trend will only continue. The value of the research is to support families in improving quality. Even though alternative care is usually compared to maternal care, the real issue is whether a child is receiving attached and committed care from one person who loves him or her dearly.

In spite of over three decades of child care wars, there is every indication—confirmed by at least three major reviews of the research—that there is a linkage between increased insecurity of attachment and extensive use of day care, especially center care, during early infancy.[17] A Canadian meta-analysis of all the studies done in Western nations since 1957 found that when young children spend more than twenty hours a week in day care, there is a negative effect on attachment, social-emotional development, and behavioral adjustment compared to children who are cared for at home. A study by Deborah Vandell and Mary Corasaniti—which the authors had a hard time even getting published— found that third-graders who had spent thirty or more hours in day care that began sometime in infancy had lower scores in work habits, emotional well-being, peer relationships and compliance. Several researchers have found correlations between nonparental care in the first year of life and more aggressive or hostile behavior in grade school children.[18] A 1996 British study found that the most positive behavioral outcomes at age ten were found among children who had had no extrafamilial child care experience.

Gunnar's findings on rising cortisol levels among children in child care settings and the strong impact on shy children generated huge controversy when they were published in 2003. Although none of the cortisol levels she recorded were high enough to be viewed as indicative of trauma, Gunnar and her colleagues held on to the results for months while they solicited comments from more than one thousand child development professionals across the world; their study was finally published in the *Journal of Child Development*.

The most frequently cited study is the National Institute of Child Health and Human Development's (NICHD) Study of Early Child Care and Youth Development (SECCYD), which has followed some 1,200 American children in ten cities across the nation. Involving twenty-eight psychologists, this study began in 1990 and is still under way. The study's major findings to date include:

Parents are the most important influence on their children's well-being, regardless of child care.

Longer hours in care negatively affect the quality of attunement and mother-infant sensitivity. More hours in care predicts less harmonious parent-infant interaction at six, fifteen, twenty-four and thirty-six months. The least securely attached infants are the babies of mothers who feel the most strongly about the benefits to their children of maternal employment. Insecurity of attachment increases when mothers are low on sensitivity, when babies are in care more than ten hours a week, and when the care is relatively lower quality or "unstable" (more than one placement).

As the hours spent in child care increase, so do rates of aggression among kindergartners, regardless of the quality of care. The more time children spend in care, the higher the rate of their problem behavior. Overall, children who spend thirty or more hours a week in after-school care have small but statistically significant increases in behavioral problems. Kindergartners who spent more than forty-five hours in care per week from ages three to fifty-four months have the most behavioral problems.[19] Seventeen percent of children who experience over thirty hours of care per week, by contrast to only 6 percent of children in care for less than ten hours a week, show aggressive behavioral problems.

The greater number of hours spent in child care prior to school, the more aggressive and disobedient children through grade six are rated by their teachers. This correlation holds true regardless of the type of child care, the quality, the family socioeconomic status or the mother's sensitivity to her child's needs. Overall about 17 percent of children have high levels of problem behavior like disobedience and "overassertiveness." A later group of researchers has broken out the rates of problem behaviors for children specifically in nonfamilial care, and they find that the rates of problematic behaviors are even higher, especially for boys.[20]

High-quality care has some cognitive benefits (at age fifty-four months.) However, cognitive outcomes do not increase with more exposure to child care, including high-quality care. Quality, not quantity, produces constructive results.

In a recent follow-up on the children at age fifteen, the effects of the quality of early child care were still evident more than ten years after the children had moved from child care into school. High-quality child care had produced "escalating positive effects" on cognitive-academic achievement. As hours of nonparental care had increased, so also had risk-taking and impulsivity. This effect was partially but not fully mediated by high-quality care.[21]

According to Jay Belsky, one of the lead researchers whose analysis continues to rankle many:

> The data . . . should encourage the expansion of parental leave, preferably paid, ideally as lengthy as it is in some Scandinavian countries, or other strategies for reducing the time children spend in non-maternal care across the infant, toddler and preschool years (e.g. part-time employment). One of the interesting questions that only history will answer is whether the cost of such leave will prove less than the consequences of its absence. Accordingly, tax policies should support families rearing infants and young children in ways that afford parents the freedom to make childrearing arrangements that they deem best for their child thereby reducing the economic coercion that necessitates many to leave the care of their children to others when they would rather not. Finally, given the clear benefits of high quality care, its expansion seems called for as well . . . these conclusions could be called for on humanitarian grounds alone.[22]

Following the reports from the NICHD study, a Melbourne University study, finding similar trends, concluded: "Children who spent more than 30 hours a week in child care centers had significantly lower social skills, were less academically able and displayed more problem behaviors than other children."[23] The Australian report was followed by two British studies, in 2003 and 2004, that again raised concern about group care for children under age two. In the 2003 study by the University of London's Institute of Education, antisocial behavior in three-year-old children was associated with high levels of group care before the age of three, and especially before the age of two. Then, in the 2004 study from the University of London and Oxford University, which followed 1,200 British children from birth to 54 months, researchers, including acclaimed child care expert Penelope Leach, research director, found increased behavioral difficulties in the older children in the study who had been in infant day care, sometimes as little as twelve hours a week. Dr. Leach, a psychologist and author of several books on child-rearing, recommends nannies or one-to-one care for youngest children and encourages parents to reserve group care for children over the age of two:

> Somewhere after two years, as the children begin to relate more to each other than to the adult, then high-quality group-based care becomes an unequivocal benefit. But for the first 18 months, all the international research shows us the importance of lots of attention from a caregiver who thinks the infant is the cat's whiskers. It may be even less important that those caring for the under-twos are trained, as that they have the right attitude toward children—that they are warm, responsive, talkative and funny.[24]

Perhaps the most telling summary of the concerns uncovered by the Oxford study was that of a British journalist who, after wading through the conclusions about the academic advantages of preschool, had the courage to say:

> It's time to be honest with ourselves and face facts. . . . The true emotional and psychological effects of neglectful or insensitive care in infancy may not show up at four or five, but 20 or 30 years down the line, in our relationships with our children, friends, husbands, and

wives . . . [in] the capacity to think, to empathize, to love . . . [in] the kind of society we're creating for the future. . . . Parents need easy access to unbiased information. An effective taboo on talking and thinking about children's needs, especially in the first year of life, is not what any of us need. On this issue at least, the time has surely come for speaking on the record.[25]

Fine-Tuning Baby Care

Babies need one person who, as Dr. Leach says, "thinks they are the cat's whiskers"; every baby needs someone who is warm, emotionally available, and attuned to him or her by virtue of interest and affection. If parents can find such a person who is willing and able to commit to their baby for a year or two, alternative care is likely to work out well. Dr. Manne advises parents to approach center care with an eye to "later entry, part time and more when children are older."[26]

Money, often the unspoken reality behind the discussion of child care quality, does not necessarily solve the problem. Certainly tuition may make huge differences in the quality of center care. But even wealthy families can overlook the emotional component in their child care choices if they believe that simply employing a nanny or another form of one-to-one care is the solution for an infant. Just take a walk on any weekday down the streets of some of the nation's wealthiest neighborhoods, like New York City's Upper East Side, and observe babies and their caregivers. You see the nannies, often in starched and pressed uniforms, pushing some of the most expensive prams or strollers money can buy. Watch for the spontaneous joy of reciprocal vocalizations or eye contact or smiles that should optimally take place between a child and caregiver out for a walk. This is an opportunity for mutual pleasure, a shared conversation or observation, engagement in a shared mood. Such exchanges can be hard to find on these streets, though they are far more important than fancy accessories. Perhaps if we reframe our search for child care for what it really is—the foundation for our child's education and health—we might be willing to spend as much time researching this decision as we spend researching and choosing our next automobile or the child's grade school or college.

Additional Implications of the Child Care Research

Another variable to consider is matching the type of care with the child's history and temperament. Was there prebirth trauma? Does the child have special needs? Has the child already endured breaks in attachment such as foster care, adoption, maternal depression or divorce? Does the child have a shy or very fearful temperament? Each of these factors can increase the need for consistent access to a warm adult who is available to the child on a committed basis for an adequate length of time.

Somewhere around the age of two, the majority of children benefit from being with other children and opportunities for learning outside the home. Spending time at a licensed home with a few children or being exposed to part-time center care is often a boon for toddlers at two and a half to three years old. For parents at home with their own children, this is the time when trips out and about with others, to parks and children's museums and playgrounds, become attractive. Parents who have been providing in-home care, drawn by their children's interests, bring them to play groups and activities like tumbling, swimming, dancing and story time so that their children can be with other children (and they can be with other parents!).

But for children who have suffered loss or profound separation from a parent, this may not yet be the best solution. Until the secure base is reestablished with an adult, such a child may be stressed by moving into peer relationships too soon. Learning how to take turns, listen, wait and share, how to not be the only little one, is a healthy stressor and important for the health of most children. Under normal circumstances, the HPA axis of children who experience such challenges returns to balance, as the system is intended to do. But if divorce, the death of a family member, a stressful move, an adoption or foster care has threatened the secure foundation of a young child, it may be necessary not to rush into exposure to long hours of peer relationships.

Child Temperament

Temperament is an important consideration in choosing child care. According to Gunnar's research, shy or fearful children can be especially challenged in a day care setting or preschool. For a child who is two and a half to three years old, it may not be separation issues that trigger the stress reaction but rather managing complex peer relations and negotiating friendships. Preschoolers care about being accepted by other children, and stress levels decrease as they gain social competence. A child does not need to be popular to maintain a low stress reaction, but overt social rejection is associated with chronic elevations in cortisol. Being able to adapt to and cope with others in a peer group setting is an important life skill that preschoolers can learn in manageable doses from adults who are tuned in to their needs and who tailor the lessons to individual temperaments.

Adverse Home Life

For children who live with a depressed or mentally ill or immature, self-involved mother, or a mother experiencing addiction or domestic violence, child care may be life-saving. IQ, social competence and emotional strength can be greatly improved by exposure to high-quality care, even for youngest children, particularly if the home is neglectful. Under these circumstances, it is especially important for alternative care to provide a warm, emotionally competent adult who can make an exceptional commitment to these babies and toddlers.

QUALITY MATTERS

While high-quality child care can be beneficial to many children, low-quality care can actually harm them.[27] In a 2010 study from Missouri of 350 children in thirty-eight licensed facilities, children were tested for social, behavioral and self-control skills, as well as for vocabulary before and after a year in child care. Because Missouri has a statewide quality rating system, the research correlated the children's outcomes with the quality of the centers. Children in high-quality programs (four or five stars) showed gains in both areas measured. But those in lower-quality programs (one or two stars) showed decreases in both categories. Children living in poverty had even stronger responses to both high- and low-quality programs, gaining more in strong programs and losing more than other children in low-quality centers.[28] Research from the 1990s indicates that the proportion of child care settings that provide good- to high-quality care in the United States is small, ranging from 9 to 14 percent.[29] Unfortunately, more recent data are not available. The worrisome reality is that we do not know the current truth about the quality of child care in our nation, but it is likely to be low for far too many children.

Although the findings on the impact of child care trigger angst, guilt, fear and sometimes deep hostility, as mothers of young children enter the workforce in record numbers, this is a topic that we can no longer afford to duck. Currently only 23 percent of American families with children under the age of six have a parent who is not in the workforce. Three out of four mothers of children under age five work more than thirty hours a week, and more than 90 percent of these families rely on some form of nonparental child care. This is a startling contrast to 1975, when only two out of every five mothers with a child under age six was employed. Women are also now going back to work sooner after the birth of a child. By 2004, 52.9 percent of mothers with a child younger than age one were in the workforce. Children under age five spent an average of thirty-six hours each week in child care; for infants, the average number of hours spent in care was twenty-nine hours a week. Thirty-four percent of children under age three and 44 percent of three- and four-year-olds also had multiple child care placements.

For many if not most mothers, there is no choice but to seek alternative care for their children as they increasingly take on the role of primary wage earner for their family. In 20 to 25 percent of dual-earner families, women are the primary source of income. In 2004, in those families where the household income was between $18,000 and $36,000, the mother's income was 66.5 percent of the total. For families where the mother was employed and the family income was less than $18,000 a year, the mother's income was typically 90 percent of the household income.[30]

Child care could in fact be a huge force for healing trauma in children and in our communities. For some families, it is that already. But for many others, it will require attention and aggressive lobbying to provide quality early education that is both emotionally and cognitively nurturing. We need to recognize that child care is the first tier of our educational system and that providers, as our children's first teachers, should be trained and paid accordingly. If our educational system is to be successful, increasing the quality of early care for all children is key to all that follows.

CHAPTER 6

Nowhere to Run

When Parents Are the Source of Trauma

MOST OF US ARE likely to have experienced at least one little trauma in childhood, usually several. Many early traumas related to surgery, short separations, divorce or even parental death can be offset—and sometimes entirely repaired—through a consistent, warm and loving relationship with one adult. But when a caregiver is an agent of fear, the child has no safe port, no predictable escape from danger, no reliable source of comfort.

In this chapter, we glimpse the "monster" unfettered within a child's home when parents cannot provide an antidote to previous and current stressors or are unpredictable—inconsistently available, emotionally and physically. When the primary agents of protection are simultaneously the agents of pain, fear takes an unremitting hold on a child's health. Although exposure to neglect and abuse has long been recognized as a potential source of trauma to children, it is surprising to learn that maternal emotional availability plays such a crucial role that, if unavailable, the toll may be greater than that of physical child abuse.

POSTNATAL MATERNAL DEPRESSION

Born wired to connect, babies have a powerful arsenal of tools to engage adults in ensuring their well-being. Baby faces, the chimerical smiles in their sleep, their intense gaze and evocative coos when they are awake, their pliable little bodies that snuggle into ours, their downy little heads that fit perfectly between our chin and shoulder—babies come well equipped to charm, given just a little time. Love in infancy is born of need, and the person who meets the baby's needs and alleviates pain and discomfort, hunger and fear, becomes imprinted in the baby's brain as synonymous for the baby with the fulfillment of needs.

As babies, we feel uncomfortable, we fuss or cry, and then something happens. To the degree that adults respond quickly and their actions are soothing, consistent and familiar, accompanied by sensations of pleasure, and match what we really need or want, we build trust and a sense of connectedness to the beings around us. Such interactions are our first experience with a sense of efficacy: "I can make things happen—when I cry, someone comes." This is also the foundation for empathy: we experience ourselves as one with "the other."

Such connections accompany attachment at any age. But they are never more meaningful than in the beginning when this picture takes form on a virginal canvas. It is ironic that our most powerful and life-shaping relationship, the one that will resonate in all the others that follow, at this stage appears so banal. Unlike romantic images of falling in love, these first "in love" experiences take place not in an art gallery or a candle-lit café, but in a bathroom, a kitchen, a nursery. These experiences don't look like swooning passion; instead, this love is manifested in tiny overlooked behaviors—a mother changing a diaper, wiping a pea-encrusted chin, settling a child into a high chair or a car seat, feeding and burping a baby, strolling with a baby in a mall. So how can this go wrong?

When Brooke Shields gave birth to her first baby girl, Rowan, she found herself overwhelmed by the contrast between how she had always imagined motherhood and the reality of her sad and disconnected feelings. Shields suffered for a while before seeking help, including medication, for her depression. After she regained her equilibrium, in the interest of mothers everywhere, Shields went public, engendering the empathy of thousands who shared her experiences of depression following childbirth. However, her disclosure that she had used a prescribed antidepressant attracted an outspoken critic in Tom Cruise, a former friend of Brooke Shields, who denigrated her concerns and denounced her for using Prozac. At this point, many professionals decided to join what had become a very public debate.

Among those professionals was Amy Timm, who wrote an article in the March 2008 issue of the newsletter published by Voices for Illinois Children. Voices is a strong and progressive children's advocacy organization that saw the debate as an opportunity to raise awareness of perinatal depression, a topic too often minimized. Timm explained that the term *postpartum depression* encompasses three types of maternal depression: the "baby blues," perinatal depression and perinatal psychosis. As many as 80 percent of new mothers get the *baby blues*, experiencing sadness, crying, worrying, mood swings and difficulty focusing and sleeping. Usually these symptoms disappear within a week or two and don't require medical attention.

Perinatal depression begins sometime during or soon after pregnancy and lasts up to a year after delivery or longer. Defined as a period that includes more than two weeks of depressed mood, ongoing sadness or the inability to be happy or enjoy life, perinatal depression is marked by fatigue, irritability, feelings of doubt, despair and worthlessness, and emotional withdrawal. Unlike the baby blues, this form of depression will *not* go away on its own. Estimates of the prevalence of perinatal depression vary. Timm estimates that it affects one in eight mothers. Other estimates range from 10 to 25 percent of all new mothers.[1] This is the form of depression we most commonly picture when we think about this issue.

Perinatal psychosis is far less common and far more serious. It may include hopelessness, hallucinations and paranoia; suicidal or infanticidal thoughts may lead to the rare but horrifying murder of a child or children by an otherwise devoted mother. This is a very unusual condition, affecting only one in a thousand mothers. Risk factors for perinatal depression or perinatal psychosis include a family history of depression or anxiety or of premenstrual mood swings. The drastic changes in hormone levels during and after pregnancy may also trigger it. Stress, including problems with a partner, death of a loved one or financial pressures, may contribute, although many women experience perinatal depression with no recognizable risk factors and may have even previously given birth without manifesting the problem.

So how and why can maternal depression cause trauma to an infant? In short, the infant's immature nervous system is easily overwhelmed, especially if the child is born with hypersensitivity from prenatal maternal depression or stress. When babies are unable to depend on a modulated adult nervous system to help them regulate their still highly labile and reactive systems, trauma can be hard to avoid. If a mother is chronically depressed after birth, she is not emotionally available to attune to her baby and "read" baby's cues. The result can be a series of tiny traumas that gradually accumulate and soon manifest as the child's "inheritance" of the mother's depression. And this is just the beginning.

Several experts are fine-tuning this picture, including Dr. T. Berry Brazelton and Dr. Ed Tronick of Boston Children's Hospital. Brazelton and Tronick began more than thirty years ago to look at how babies' behavior with their mothers is affected by even a mild interruption in the dance between them. They asked mothers to bring their approximately four-month-old babies into the labs at a time when the babies were not normally napping or hungry. Two cameras—one focused on the baby, one on the mother—projected simultaneously on a split screen to give close-up views of the minute facial expressions and physical cues that each mother and baby were signaling to each other. They also measured heart rates for both babies and mothers and the conductance of moisture on the palms of their hands.

The babies were awake and alert in infant seats, facing their mothers only a few inches away. Mothers were instructed to play with their babies. The cameras captured reciprocal smiling, talking and cooing; as mothers' hands reached out to play pat-a-cake or other games, touching and caressing, babies' arms and legs moved rhythmically toward their mother's face and hands. Babies looked directly toward their mother's eyes to engage, though occasionally they would break mutual gaze by turning their heads away, putting their fists in their mouths, and looking at a proffered toy or something else. Typically mothers responded to this regrouping of attention by allowing the baby to set the pace, commenting on whatever had attracted baby's attention, and reengaging when the baby was ready.

Next the researchers directed each mother to become like a statue and to not play or smile or respond to the baby's efforts to engage her, simulating depression. At first most of the babies were puzzled: their brows knit in confusion, and some worked hard to vocalize and kick and smile and laugh to get the mother back again. As she stayed frozen after several tries, amazing behavior began to emerge from the babies. At first they tried to repeat behaviors that had worked before—vocalizing, smiling, raising their eyebrows, pointing, pouring out all of their charming repertoire, while carefully watching their mother's face to see if it was working. Within seconds, most babies began to fuss and attempt to get their mother's attention by ramping up their distress cues to full-blown crying and flailing. As the mothers remained disengaged, some of the babies even risked their own stability, kicking themselves forward, arching their backs, nearly propelling themselves out of their infant seats in their efforts to get her attention. When the researchers cued the mothers to reengage with their babies, the babies' responses were again surprising. Some were eager and responsive, quickly returning to the way they usually interacted. A few, however, were still unhappy with their mother and either wanted more profound reassurance through holding and intense attention or turned away—as if punishing their mother!

For most caregiver-baby dyads, the regulatory rhythms of one brain actively mold those of the other. Ideally the adult is leading the dance. An everyday example of the two-way nature of this dance is feeding. When baby is hungry, he cries. Hearing baby cry, mother's milk lets down, and she is ready to nurse. If the baby is still asleep, she may rouse him to relieve her discomfort and in so doing begins to regulate the baby's rhythms to match her own. How much milk the baby takes in over time controls how much milk the mother produces; this is a cycle of mutual reward.

Through everyday interactions, including feeding, holding, changing and soothing to sleep, the adult helps the baby regulate physical and emotional states by anticipating the baby's needs, "reading" the baby's cues and responding constructively. When baby is highly aroused, crying and upset, mother comforts and soothes by holding, rocking, patting, rubbing a little back and using her voice: "There, there, you're okay. Everything will be all right." If baby is sleepy in the early evening but mother has learned that keeping her sleepy baby awake in the early evening enables the baby to sleep through the night, she learns to arouse her baby, raising her up over a shoulder, bouncing her, tickling little feet, and talking in a high-pitched excited voice to wake the baby up. The baby's responsiveness to her techniques for rousing or soothing, burping or feeding, shapes what the mother repeats.

Sleep-wake states, heart rate rhythms, eating and digesting rhythms, and reciprocal vocal and visual communications will gradually begin to mirror each other's, so that ideally the rhythm between the two will usually come to resemble a waltz. If the adult's system is calm and well regulated, barring serious pathology, the baby's system will gradually reflect that adult's healthy rhythms. If the adult is anxious or depressed or her life is erratic and chaotic, the baby's patterns will soon follow her lead so that the dance between them will look less like a waltz than a rumba.

To the onlooker, maternal depression may appear contagious to her baby. But this likely stems more from nurture than from nature. Let's look more closely. Awake and already fed, baby looks to his mother with raised eyebrows, gazing at her face. Encased in her own sadness, mother misses baby's cue and looks away. Or vice versa: mother looks to the baby's face, smiling and reaching, and baby, overwhelmed or tired, arches his back and looks away. The process of connecting between babies and caregivers is characterized by thousands of tiny cues, some of which are read accurately and attended to with mutual satisfaction, while others are missed or misread. Tronick says, "There is disorganization, mismatches of affect and intention, error and sloppiness."[2]

But here's the key: mismatch or "messiness" can be repaired by adult regulation of the infant's system as they continue to interact. The opportunity for repair is ever-present in all of the tiny interactions that happen throughout each day. Between infants and adults, as between two adults, mismatches in relationship tend to create a negative mood for both participants. But they have an opportunity to rematch their emotional cues and to respond accordingly. So when a responsive mother tries to get baby to look at her and baby looks away, she allows the baby to disengage and waits for him to show interest, through a look back at her, before trying again. A depressed mother is more likely to feel hurt by the baby's turning away—a "rejection" that validates her sense of failure—and she may turn away herself. Or she may try to reassure herself of her baby's love and try invasively to re-engage the baby. Baby, feeling his mother's anxiety, looks away again. Over time the baby becomes a poor interactive partner. Sealed into her own world without awareness of the baby's state, a depressive mother tends to interpret her baby's avoidance of her attempts to engage as "my baby doesn't love me." Her interpretation contributes to her depression and consequent withdrawal or intrusiveness with her baby. Compounding this cycle is the reluctance of most depressed mothers to openly discuss these feelings.[3]

Dr. Tiffany Field, director of the Touch Institute in Miami, finds that infants of depressed mothers carry their negative pattern into relationships with new adults. These babies neither expect nor have learned how to engage another person socially. They are likely to seem "blah" or disinterested and to look away from the face of the person holding them, gazing around the room, where they have learned to seek stimulation rather than from their "dead" mother.[4] Dr. Tronick finds that this response has a strong dampening effect on even highly nurturing adults. When the adults in his lab, who were not told which infants had depressed mothers, tried to interact as warmly with these infants as they normally would with any infant, they found themselves doing less touching and smiling and interacting less frequently with the babies of depressed mothers. The babies' detachment had already had a profound impact on a new relationship.

According to Dr. Tronick, the interactions of depressed mothers with their babies fall into two categories: some mothers are intrusive, speaking in an angry tone, poking the baby, being rough, and often interfering with the baby's actions; others are withdrawn and disengaged, doing little to support or be involved in their infant's activities. In Tronick's study, the babies in each of these two groups were correspondingly different. Infants of intrusive mothers kept looking away from their mother and seldom cried, though they expressed anger. Infants of withdrawn mothers cried and were more overtly distressed or sad. The differences between the two kinds of depressed mothers correspond with differences between babies who are abused and those who are neglected. Babies who are parented by intrusive mothers or who are abused develop "an angry and protective style of coping" in anticipation of the mother's intrusiveness, tuning out or pushing away.[5] At least there are intervals of reparation for these infants between intrusive or abusive episodes. Although they become more easily frustrated or angry and are often hypervigilant, the babies of these mothers are still able to engage with others. Their anger may actually afford some escape from the mother's imposed pattern. But the babies of disengaged or neglectful mothers are left alone to self-regulate and receive little or no respite from their overwhelmed systems. They are likely to withdraw and may even "fail to thrive," as was the case for many of the children in Romanian orphanages. It appears that baby boys are particularly vulnerable.

When postpartum depression lasts more than a couple of weeks, it tends to become a chronic state. Without intervention, maternal depression scores remain stable over the course of the first year postpartum.[6] So symptoms are not transient—they do not come and go. Babies of depressed mothers are interacting within a consistently deprived relational dance where there is little smiling, laughter, facial animation or delight. Seriously depressed mothers are also likely to have other mental health diagnoses, including anxiety, often in combination with poorer self-esteem and less confidence in their own mothering.

From the baby's perspective, the issue is the unavailability of the person they most want to engage. Attachment, the engagement of a committed adult to feed and protect the baby, is essential for survival and is the first major task of infancy. This connection is foundational to the child's regulation of emotional and physical states and emotional security, including a positive sense of self and of personal efficacy. Most babies are successful in this vital task. But the adult leads the dance. And if the adult is robbed of vital energy by depression that impedes responsiveness to an infant, the dance of attachment is impaired. Baby may be left at the mercy of his or her vulnerable nervous system, swept under by the tides of unregulated needs and emotional states without the benefit of the mother's external modulation. By two months of age, infants of depressed mothers often have difficulty engaging in social interactions. They are fussier, less able to regulate their emotions, and less physically active than other infants. By three months, their EEG patterns are similar to those of adults with depression. By age one, they are behind on developmental tests, and at age three they have lower scores on both cognitive and linguistic measures. By the age of four, and again at eleven, children of depressed mothers have lower cognitive scores on developmental tests compared to other children their age. In addition, by their first birthday these children are significantly more likely to be insecurely attached, another risk factor for increased vulnerability to trauma.[7]

Depending on its severity, many researchers regard maternal depression as an experience as damaging as child abuse. Some refer to this experience from the baby's perspective as "the dead mother" and point to the fact that the child ultimately regards himself as not important enough to enlist his mother's attention and love. There are observable early differences between these babies and babies of nondepressed mothers. One notices less eye contact and less smiling. Babies are drowsier, more passive, or more temperamentally difficult. They are often more anxious and less able to tolerate separation. Both mothers and babies have a more difficult time engaging in shared emotional states, and one will often trigger a negative mood in the other. From the baby's perspective, he or she is failing to experience pleasure with or from another, dampening the baby's perception of reward in human interaction and leading to less confidence in future relationships.

Unlike a nondepressed mother, a depressed mother is often inconsistent—sometimes withdrawn, sometimes nurturing. It's hard for baby to anticipate how the mother will be or to relax in the confidence that she or he knows what to expect. A depressed mother may also be slow to respond to the baby's efforts to engage her vocally and fail to initiate vocal exchanges that encourage language development. And when they do talk to their babies, depressed mothers don't sound like regular mothers: the lilting, joyful vocal tones of "motherese" and the fun of playing "pat-a-cake" and "peek-a-boo" or mimicking funny faces are mostly missing. Modulating the tone and tempo of interactions that help the baby regulate his or her behaviors may be hard for a depressed mother, so that, for example, when her baby is overexcited, she may not know to gentle her tone and lower the baby's state to avoid a meltdown.

All of this can be headed off if mothers are screened, assessed and treated for depression or if they receive early attachment-focused interventions. One highly effective example of how to do this is the Nurse-Family Partnership (NFP) program. NFP is the gold standard for home visitation of newborns, both in the United States and in Britain. The program targets vulnerable parents of firstborn babies, especially teens and parents experiencing risks such as poverty or mental or physical illness. Beginning during pregnancy and lasting until the baby's second birthday, specially trained nurses go to families' homes to answer questions and help prevent or temper the foreseeable challenges to earliest development, including maternal depression, parental immaturity, domestic violence, addiction and mental illness. Rather than waiting for predictable problems to arise that will take their toll on the young brain, visiting nurses provide parents with information, counseling and linkages to resources that scaffold them. An investment in zero-to-three education of all children, NFP, where it exists, fills a huge gap in current social services, which are primarily designed to repair the same children later at great expense and with less effectiveness. Community-based screenings for perinatal depression built into the NFP model provide an invaluable strategy for reversing and preventing generational transfer of this issue (see appendix E for a summary of NFP outcomes).

Symptoms of early trauma may be hard to distinguish from the normal developmental stages of toddlerhood—like the "terrible twos" (which, by the way, are frequently the "terrible threes" as well!). Even after a child has witnessed a violent murder or suffered a grievous loss, many people will say that the child seems "fine," because she or he is passive and obedient. We tend to assume that nothing major has taken place, and as concerned parents we want to believe this is true, but the reality is that because children can neither fight nor flee, passive and obedient behavior may simply be their only option. The effects of trauma may not show up until much later, when they manifest in abnormal behavior that only seems to come out of the blue. Or the effects may remain dormant far longer; the impact on a child who appeared resilient and unaffected by trauma may surface years later in relationships with other people or in the form of a health diagnosis like hypertension, diabetes or a chronic digestive disorder. "Children aren't resilient," says Dr. Bruce Perry. "Children are malleable."

As a family therapist, I can testify to the generational repercussions of maternal depression in intimate relationships between couples. For example, one wife complained in therapy that her husband didn't pay attention to her emotionally, was not available for close conversations and affectionate exchanges, and went right for sex when her desire was for intimacy. When she asked, explained or begged him for a bridge of sensual communication leading up to sex, he would seem "removed" and distracted, she said, and at the next opportunity for intimacy it was as though he had never heard her request. I interviewed each partner about their families of origin, and it soon became clear to me that the husband had been raised by a mother who gave him and his siblings a strong start educationally but was unavailable to them for any sort of warm, sensual exchanges—affectionate play, nuzzling, hugging, caressing. His mother had given her children the message that there was strength in being "independent" and staying cool and reserved; she seemed not to have valued "the mushy stuff." This had been the legacy of her own childhood with a depressed mother and an absent father whose central value was academic achievement.

The couple had two small children whom his mother occasionally babysat. The husband was particularly in love with his firstborn son, a five-year-old who looked very much like him. I asked the husband to observe his mother with his son and to note how she interacted with the little boy—a five-year-old version of himself. I gave him specific behaviors to look for that would capture the connections that his wife had identified as missing between them and that were in place with his son, who actively sought them because he enjoyed them with his emotionally nurturing mother. So the husband stayed one day while his still chronically depressed mother kept the children; he lingered in the background and watched for behaviors like joyful exclamations at his son's successes, the availability of fantasy or any shared special routine between his mother and his son, hugs, caresses, warm voice, or affection shown in any nonfunctional way (a good-bye kiss, for instance, didn't count). He saw, and felt through his son, the missing warmth and vivacity in his mother's interaction with her grandson. This was the trigger he needed to open a long-closed door in his heart, and the marital work shifted into gear. And physical healing may not have been far behind. The husband came from a long line of men who had died early deaths from cardiovascular disease. I can't help but believe that this man's "change of heart" was just that.

In summary, maternal postpartum depression should not be dismissed as a benign and normal condition. Far too prevalent, it conveys grave risks to a child's health. Left untreated, the impact on infants can be as profound as child abuse, if not more so. The critical issue is the chronicity of the mother's depression and the accompanying detachment. An early study in Minnesota looked at the effects of different kinds of maltreatment.[8] Forty-four abused babies were divided into four groups: those who had been physically abused, those who had been neglected, those who had been rejected with hostility, and those with psychologically unavailable mothers. At eighteen months, the infants of the psychologically unavailable mothers, who were unresponsive, detached, depressed and uninvolved, were the most adversely affected of the four groups of babies. Though their mothers had not mistreated them in any other way, they had shown little pleasure in interacting with their babies and had failed to comfort them when they were upset. From nine to eighteen months, these children declined the most rapidly on the Bayley Scales and showed the most severe and varied problems of all the babies. As infants they failed to develop trust, and as toddlers they failed to negotiate autonomy. In a review of the literature, Dr. Frances Thompson Salo of the Murdoch Children's Research Institute concluded: "Our understanding is that the child responds to feeling that he is not in his mother's mind and could therefore get lost by feeling extremely threatened. If unrelieved, this state can be traumatic." Dr. Thompson Salo is also convinced that there is a linkage between the experience of such early depression and ADHD, which shares many symptoms in common with PTSD.[9]

Chronic maternal depression is also often at the root of child neglect. Behind the unkempt house or the poorly clothed or fed child, or a baby failing to thrive, may lie maternal depression. And behind mother's depression often lie unspoken realities, such as domestic violence, a risk of maternal suicide, eating disorders or addiction.

The key to prevention and intervention is seizing natural opportunities to screen mothers and babies. Dr. Brazelton views routine exams by obstetricians, midwives, pediatricians and other health professionals as potential "touchpoints." His Touchpoints Program trains health practitioners across the country to use health visits and the child's developmental milestones—the language of the child's behavior—as opportunities to intervene early where maternal depression, or any other challenge to development, is suspected.

Before leaving this topic, we should note that almost all of the research on the impact of depression on the developing brain to date is on the maternal-child relationship. Paternal depression and mental illness, however, can present equivalent threat to a mother and consequently to a baby; if he is the primary caregiver, similar dynamics pose equivalent risks to a baby. Reliant primarily on animal studies, in which the female is entirely responsible for care of the young, research has not yet caught up with the changes in human parenting practices whereby fathers as well as mothers may be the primary caretakers of young children. Children's health is a societal and particularly a familial responsibility, and the risks to infant development from maternal depression should not be blamed on mothers. The dynamics set in place by paternal mental illness or drug abuse or by domestic violence can exact at least as large a toll on maternal and child health and may be a significant contributor to maternal depression and the dynamics described earlier.

One positive note is that although babies of depressed mothers may generalize their negative behavior patterns to their relationships with other adult women, they usually do not behave in the same way with their father if the father has been present from early in their lives. When a father or another consistent alternative nurturer (such as a grandmother) is regularly available to the baby within the family, and is not depressed, their interactions with the baby may help buffer the effects of maternal depression.

PARENTAL ABUSE AND NEGLECT

The most insidious source of trauma to a child is when something that should be safe is not. Parents and key adults in a child's life should be sources of nurture and comfort, but for far too many children the adults in their environment are a source of suffering as well as the only form of nurture they have known. Among the rich democracies, the U.S. child abuse death rate is nearly double that of France, three times higher than Canada's, and eleven times higher than Italy's.[10] The younger the child, the greater the risk. In this country, 85 percent of fatalities from abuse and half of all victims of abuse and neglect are children under six. Abuse is among the leading causes of death in the first year of life.[11]

Domestic violence in this country is widespread: at some point in their lives, 25 to 31 percent of American women are physically or sexually abused by a husband or boyfriend, including hitting, kicking or stabbing. Pregnant women are frequent targets of such abuse, which magnifies their risk for preterm labor and delivering low-birth-weight babies.[12] From 3 to 10 million children witness such violence each year and are more than fifteen times more likely to be abused than nonexposed children.

In the most prosperous country in the world, child neglect is even more pervasive than child abuse. More than 500,000 children are reported for neglect each year, double the number reported for physical and sexual abuse combined.[13] Child abuse and neglect are the Pandora's box of linkages between early trauma and adult ill health. Though the research has been prolific, the effects are just beginning to be understood.

While "abuse" and "neglect" are umbrella terms that encompass a range of different kinds of experiences, the common denominator is the toll of trauma on a developing brain. Whether a child is hit, shaken, beaten, exposed to watching this happen to someone else, or neglected, his or her HPA axis is being highly stimulated. In cases of neglect, the added element of deprivation of emotional connections in the brain makes this a particularly devastating form of trauma, and one hard to recover from. In both abuse and neglect, lack of safety in a child's world undermines normal development. But in neglect, particularly in cases of global neglect (such as the feral children Dr. Bruce Perry writes about in *The Boy Who Was Raised as a Dog*), the child's brain is deprived of the nutrients of life. Especially in the beginning, if a baby is not held, touched, rocked, comforted, talked to and played with, the neurons waiting for stimulation don't connect.[14] It's as if the seeds are all there but there is no soil, no sun, no water. Without nurturing interactions, there is no growth. In extreme cases, babies die.

Messages from the adult's right brain to the baby's right brain—the nonverbal, gestural and tonal communications—mold a developing nervous system. Our ability to calm after a frightening event evolves from someone who once responded to our fear, picked us up, and provided a shoulder, rhythmic pats and the dance of comfort to soothe our overwhelmed immature self. We depend on the mature nervous system of an adult to save us from the open seas of our own unbridled emotions in the beginning of life. One system shapes the other, and these early patterns become foundational. Harlow's monkeys gave us a searing picture of the importance of touch and movement: even though the wire surrogate mother was equipped with food, the babies preferred the cloth mother when frightened. In subsequent research, little monkeys clung to a swinging wire mother over a stationary wire mother to obtain the vestibular stimulation that is as essential for monkey brain growth as rocking is to baby humans. Without appropriate stimulation, brain circuitry doesn't proliferate, either in little monkeys or in little humans, particularly in the right orbitofrontal cortex of the brain; it is next to impossible to recover these capacities later, at least to anything approaching full potential.

Whether a baby is left to cry alone or is frightened, she is facing conditions that overwhelm her capacity to cope. Think about a time when you were really frightened as an adult. Remember what happened to your breathing, your heart rate, your stomach, the palms of your hands. A baby has the same reactions to fear: the fight-or-flight response kicks in, and heart rate, blood pressure, respirations and muscle tone all increase. The child becomes hyper-alert to the threat, so that all information not related to the perceived danger is tuned out in her awareness. Norepinephrine is released, and all regions of her brain that regulate arousal in the lower brain are turned on. Since she is an infant, she will use her entire repertoire of behaviors to bring the caregiver to her: her facial expressions and body movements will change, and she will probably cry to call for help. When this works as it should, the caregiver comes to pick her up, soothe, feed and comfort her, and bring the child back into a modulated state.

RELATIONAL TRAUMA

But if no one responds to the baby's cries and she is left to fend for herself, or if she is abused or exposed to violence by the hands that should be there to rock the cradle, her lower brain will be highly stimulated. Because the brain builds itself from experience, this kind of stimulation, if it occurs chronically, will build an overactive or highly sensitized response system in this little brain. Now one of two things will take place, depending on the child's age, circumstances and temperament. She may remain in a state of hyper-arousal, crying and flailing about. Or, overwhelmed by terror, she may, like an electrical circuit receiving too strong a current, simply switch off—she freezes and dissociates. In extreme cases, she may faint. For many children, this is the beginning of "learned helplessness"—going directly into passive defeat in the face of threat rather than putting up a fight.

When these infants are boys, they tend to become aggressive, impulsive and reactive. In a persistent state of low-level fear and anxiety, the part of their brains that processes their world is different from the part of the brain employed in children who experience themselves as safe.[15] As early as preschool age, they may become violent "out of the blue"—with seemingly little provocation. When they are girls, they tend to dissociate, appearing distracted or not there. Because in young children the cortical or "thinking" brain is only a thin layer early in life and develops more slowly than any other part of the brain, the memory of early trauma is stored in the lower brain, typically the limbic brain, as a somatic or emotional memory, where it remains inaccessible through language or rational thought when these abilities develop.

When the parent is the source of pain, either to the child directly or to someone the child loves, comfort is paired with hurt—the object of desire is also the instrument of agony. This is the birth of ambivalence. For anyone who has wondered how an adult can be attracted to a partner who beats or otherwise hurts him or her, look for the possibility of an early history of such experiences. First relationships are the prototype for future relationships. In an unconscious effort to resolve the trauma, many people gravitate to people and situations similar to the unresolved early dynamics.

When the parent is unavailable, the child remains in a state of incomplete emotional development, experiencing a fundamental sense of failure. Capturing the love of an adult is after all the baby's first task. Though recorded with neither words nor rational thought, the experience is: *No*

matter what I do, all my best efforts to bring her to me and to keep her atten-
tion are not enough. . . . I can't. I'm not enough. . . . She is elsewhere. . . . She
isn't with me. I can't. Again, such transactions with our first love may shape
relational dynamics we re-create in future relationships.

In early March 2009, Oprah Winfrey devoted her show to serious
neglect in children. She featured a little girl named Danielle, who had
been found in a filthy house in a back room where she wore a diaper and
drank from a bottle. Danielle could not walk or talk, though she was
more than six years old. She weighed forty-three pounds and had the developmental skills of a six-month-old. Her mother had provided her with
milk in a bottle and not much else. Confined to an empty room with
linoleum floors, Danielle slept in a box and crawled daily through her
own wastes, mostly unclothed. Her mother had kept Danielle's birth and
presence in the home secret until she was two. When neighbors saw a
child appear and quickly disappear at a window one day, they called child
welfare. But when a social worker made a call at the home, she neglected
to wake up the child when the mother said she was sleeping in a back
room safe and sound. Danielle was left in these desperately neglectful
conditions for another four years before she was finally rescued. When
she was removed and later adopted by exquisitely patient adoptive parents, they said that when she first began walking, she would constantly
fall down and hurt herself. But she didn't react to the pain of a skinned
knee or other obvious injuries. Without realizing it, Danielle's parents
were perfectly describing the combination of learned helplessness and
dissociation typical of children who have suffered early constant trauma
from maltreatment.[16] In spite of world-class adoptive parents, Danielle is
living testimony to the fact that there are irreversible losses from profound maltreatment early in life.

Harlow's baby monkeys provide another poignant snapshot of the effects of neglect. These little monkeys who were separated from their
mothers and isolated or given minimal contact with a nurturing surrogate
were the picture of dejection and anxiety. They were reluctant to explore
their environment, and their learning was impaired. They were unable to
integrate competently into a group of their peers. Their emotional and
physical health was severely compromised. Subsequent to Harlow's original work, his associate Dr. Steven Suomi found that little monkeys

who were essentially raised by each other became adults with impaired impulse control, excessive alcohol consumption, anxiety, social withdrawal and impaired function of the stress axis.[17]

Dozens of studies have looked at the clinical implications of abuse and neglect (see appendix B). Most of the studies on maltreatment have been retrospective in their design, like the ACE Study. Several current studies are following children forward as they develop, including the U.S.-based Longscan Study, which began tracking maltreated children in 1990, from age four to adulthood; as the children mature, Longscan will provide powerful information on adult health.[18]

For a prospective study of physical health that begins at birth, we turn to the Dunedin Study in New Zealand, named for the place where the children were born and most of the subjects still live. Researchers are following more than one thousand children born in 1972 and 1973. The study defines child maltreatment as rejection by the mother, frequent changes of primary caregivers, physical abuse resulting in injury, and sexual abuse during the first decade of life. Now thirty-five and thirty-six years old, participants were most recently assessed in 2005 at age thirty-two. Because the group is too young to have manifested heart disease, the study measures inflammation as a precursor to cardiovascular symptoms. Even mild levels of inflammation predict increased risks of heart disease, as measured by blood levels of C-reactive protein, fibrinogen and white blood cells. The study controls for the influence of medications and differences in income levels as well as the presence or absence of a healthy lifestyle.

The Dunedin findings reveal that maltreated children suffer both psychological and physical effects well into their adult lives. Though they still appeared healthy, adult survivors of childhood maltreatment were in fact more than twice as likely as their nonmaltreated peers to show clinically relevant inflammation levels twenty years later. Inflammation not only is linked to increased risk of heart attack and stroke but also predates the development of hardening of the arteries and chronic lung disease.[19] This research promises a new window on the ACE correlations, one that may lead to treatment of inflammation and other key warning signs earlier in development.

RISKY FAMILIES

Another area of study, known as "risky family" research, attempts to further delineate families most at risk of maltreating their children. Although there are no perfect families, there is a strong relationship between the number of risk factors that challenge family safety and the emotional and physical health of the family's children. From a global perspective, conditions like war, environmental toxins, disease and even natural threats like volcanoes, fires and floods have an influence on family health. Within a culture, we have known that social and political factors surrounding families clearly play a role, as do status and wealth—or the privation of these. In our own society, low socioeconomic status (SES) alone is associated with more irritable, punitive and coercive parenting, greater likelihood of exposure to violence, increased rates of alcohol or drug abuse, exposure to secondary smoke and a fattier diet. In the United States, poverty is associated with lower educational achievement, higher rates of arrest, lower income level, higher divorce rates and lower occupational status, each of which exacerbates the likelihood of negative health outcomes.[20]

As we fine-tune our lens to look more closely at the dynamics that tend to be learned within families, we can readily see that the most serious risks to child health are harsh or coercive parenting behaviors. The problem with these kinds of parenting behaviors is that they fail to teach the child and may in fact impede the child's ability to learn constructive self-regulation. This critical skill can only be learned in a healthy environment—which fundamentally means a safe environment. Healthy families provide physical safety and emotional security, and they teach behaviors to foster their children's independent and constructive decision-making so that they can eventually maintain their own physical and emotional health. When parents fail to teach and model skills that enable children to internally manage their reactions to life's inevitable stressors, their children's stress response systems will be chronically stimulated and slower to return to balance than those of children who have learned self-regulation. As a result, children's immune and endocrine regulation is jeopardized. The quality of early family life and

of parenting dynamics has a huge impact on how this alarm management system develops.

The ACE Study delineated familial risk characteristics like parental
mental illness, addiction and incarceration. The "risky family" research
looks more closely at a clearly destructive set of characteristics—often
seen in everyday families—that are typically overlooked in the studies of
child maltreatment. The goal of this research is to determine how relatively typical family-of-origin dysfunction takes a toll on children's health
long after it occurs. Although these behaviors are subclinical—that is,
they are not generally diagnosed as pathological—they do in fact produce more modest versions of the same types of pathology in children
that are commonly associated with child abuse or neglect. From these
studies we now recognize, for example, that "overt family conflict" (recurrent episodes of anger and aggression) and what is called "deficient
nurturing" (interactions with children are cold, unsupportive and deri-
sive) result in parental failure to impart constructive self-regulatory skills
to children. Such patterns often pave the way to an array of emotional,
physical and behavioral disorders. As was also observed in the ACE Study,
researchers looking closely at these families found that their children were
much more likely than their peers to turn to smoking, alcohol and drug
abuse in pre-adolescence and in the teen years, decisions that in turn
compound both emotional and physical disease.

For young children, "overt conflict"—anger, hostility and aggression
toward the child or between the adults in the home—is toxic to health.
Overt conflict describes a wide range of destructive scenarios, from living
with irritable and quarreling parents to witnessing violence to being
abused. Higher levels of coercive control in a family correlate with a lack
of warmth, acceptance and support—the basic food for emotional
growth. The result is often a generational legacy of aggressive and noncompliant behavior in children. High levels of conflict in a family correlate with conduct disorder, delinquency, antisocial behavior, anxiety,
depression and suicide in adulthood. "Deficient nurturing" covers a range
of family characteristics, including emotional neglect, unresponsive or
rejecting parenting, lack of parental availability, lack of involvement in
the supervision of children and detachment or alienation on the part of
the children.[21]

It is important to note that these risk factors are not limited to underclass families, as evidenced by the ACE research, which studied a middle-class and primarily white population. In extreme cases, both permanent disability and death can result from "risky family" behaviors, and a range of physical diagnoses are associated with these experiences. In addition to the correlations with ischemic heart disease, obesity, chronic lung disease, skeletal fractures and liver disease found by Vincent Felitti and Robert Anda, other studies have linked risky families to poorer growth in infancy and lower height attainment at age seven. As adults, those who grew up in risky families had higher rates of minor infectious disease and increased risk of some forms of cancer.[22]

Collectively, the "risky family" studies point to the need for early intervention. So how do we identify these families? The maternity nurse is a good place to start. Although some factors can be hidden, like drug abuse, alcoholism or emotional abuse, the majority of risky families are recognizable at the time of birth. Maternity nurses know which babies are going home to risky families: teen mothers without support; babies born addicted to or affected by alcohol or drugs; and babies whose mothers are too overwhelmed to count little fingers and toes, have bruises on their faces or backs, are reliant on medications for chronic mental illnesses, or have no one with them who is thrilled at the birth of a new family member—just to name a few signs. Somewhat harder to detect are mothers who are quietly submerged in angry or demeaning relationships or who display few signs of joy or pleasure in their baby. Medical practitioners tend not to ask questions. Instead, we wait for little battered bodies to appear with the full-blown signs of trauma, often too late to protect the potential with which they came into the world.

THE IMPACT OF EARLY MALTREATMENT ON MENTAL HEALTH

As recently as the early 1990s, health professionals still believed that emotional and social pathologies were the result of the relatively ephemeral notion of "psychological dysfunction." Mental illnesses such as schizophrenia and bipolar disorder were seen as primarily genetic in origin, best treated by chemicals. Just twenty years ago, the role of brain-based dysregulation in, for example, personality disorders was not yet recognized.

In the 1980s, Dr. Martin Teicher, director of the Developmental Biopsychiatry Research Program at McLean Hospital in Belmont, Massachusetts, had begun to suspect that something else was at work in the symptoms he was seeing in his borderline patients. Teicher's research led him to question the function of the limbic brain, especially the amygdala and the hippocampus, both of which play a strong role in emotion and memory. Teicher suspected that early abuse was at the root of changes in the structure and function of these areas of the limbic brain. Drawing upon a sample of more than 250 subjects, he and his colleagues designed a questionnaire to determine whether these two areas of the brain could be damaged from early abuse by excessive exposure to stress hormones. Their results confirmed their suspicions.[23]

Teicher found that exposure to glucocorticoids (cortisol) from maltreatment at critical periods of early development not only changes brain function but also has an "organizing effect" on development. The impact on mental health is profound. For example, post-traumatic stress disorder and dissociation are actually adaptations that enable us to survive extreme maltreatment in a relatively insulated state, emotionally removed from pain. Similarly, anxiety enables a child born into a strife-ridden world to be alert to the next source of threat.

According to Teicher, a central issue here is timing.[24] The McLean study found that when it comes to mental health, early emotional abuse is even more likely to produce psychiatric symptoms in adults than physical abuse. Teicher suggests that "combined exposure to less blatant forms of abuse may be just as deleterious as the most egregious acts we confront."[25] Critics have questioned the validity of self-reported experiences, in both this study and the ACE Study. But Teicher and his colleagues cite research concluding that adult self-reports actually minimize rather than exaggerate adverse childhood experiences, particularly verbal abuse. Another study found that 63 percent of parents in the United States report one or more instances of verbal aggression toward their children, such as swearing, belittling or insulting.[26] Child abuse rarely occurs in neat categories; for example, physical abuse is most often accompanied by emotional and verbal abuse and neglect. Victims in this study remembered the pain of verbal and emotional aggression more than the physical abuse—which, ironically, was probably the only catalyst that would trigger the involvement of state authorities.

In a second study, Teicher drew upon the childhoods of 554 adults. This study found that the highest correlations were between maternal verbal abuse and a heightened risk of personality disorders, including borderline, narcissistic, obsessive-compulsive and paranoid personality disorders. This finding is surprising considering that child welfare officials overlook this form of abuse entirely. In response to an overwhelming number of children in need of protective services, current state and federal priorities are focused almost entirely on the impact of physical and sexual abuse, which may actually have fewer long-term negative consequences.[27]

The most serious pathology seen in this study was in individuals who had experienced multiple types of abuse. As was true in the ACE Study, Teicher relied on questionnaires completed retrospectively by adults who had already been diagnosed with various pathologies. A major difference between the studies was Teicher's focus on mental rather than physical health. His subjects were predominantly white (73 percent), younger than the ACE subjects (ages eighteen to twenty-two), and predominantly female. Additional correlations between early abuse and mental health included the following:

Dissociation: Dissociation interferes with normal learning and with commonsense judgment, especially in relationships. Emotional abuse is more likely than either physical or sexual abuse to generate dissociative experiences in young adults. Verbal abuse and witnessing domestic violence are each moderately correlated with the symptom of dissociation. The most surprising finding was that children who had both experienced verbal abuse and witnessed domestic violence were even more likely to suffer from dissociation than children who had been sexually abused.

Anxiety, depression and anger/hostility: A combination of exposure to verbal abuse and witnessing domestic violence is as great as, or greater than, the impact of familial sex abuse in producing symptoms of anxiety, depression and hostility. The effect of this toxic combination is equivalent to the effect of experiencing physical, emotional and sexual abuse combined.

Substance abuse: Early maltreatment is clearly implicated in the later development of substance abuse and is also linked to depression and anger/hostility.

In summary, the destructive impact of domestic violence on health does not simply appear out of the blue in adulthood. For those who are looking, the connections are visible much earlier. Research on 160 preschoolers from low-income families in Michigan revealed that children exposed to violence already had symptoms of PTSD such as bedwetting, nightmares, allergies, headaches and greater susceptibility to flu. These children had four times the rate of asthma and gastrointestinal problems as their peers. Both being abused and having a mother who abuses substances are correlated with health problems for young children. The mother's poor health and the child's level of trauma are the strongest predictors of the child's poor health. A particularly alarming finding in this study was that nearly four-fifths (78 percent) of the children had been exposed to some kind of violence, within either the home or the community, and that nearly half (48 percent) had been exposed to at least one incident of mild or severe violence in their own family.[28]

The research reveals that our core emotional programming develops in layers, like a tree. If we cut a cross-section of the trunk of an older tree, we see that the tree, like the earth's surface, grew in layers. If there was damage to the tender sapling, subsequent layers grew around the abrasion. From the outside, the bark looks like that of every other tree, but when we look inside we see the wave formed as each layer wrapped around the early wound. This pattern is also true for people. A child who is smart and athletic and looks like any other child on the outside may be extraordinarily competent and grow up to be a Harvard grad, a judge or a physician, but he may also be the first among his peers to suffer a heart attack, joint or muscle disease, or gastrointestinal disorders. Families may be right when they say, "Heart problems run in our family," but the trigger may be something other than genetics. Or the child grows into an adult who, though extraordinarily competent and successful, outgoing and confident, may approach every aspect of her life with anxiety and perfectionism, so driven that any mention of concern is heard as a sign of weakness. Her health will be affected, if it is not already.

Each decade adds layers around the original traumas of our early lives and renders us relatively more vulnerable to later stress, especially in the area of the original injury. Later traumas may also take their toll, but how we react to them will have everything to do with the cumulative foundation of earlier experiences.

When first love is erratic and unreliable or inseparable from pain, our expectations of other loves and our sense of self in relation to the other are likely to be skewed accordingly. This is the legacy of neglect or abuse and of insecure or anxious or ambivalent attachment. These experiences both disallow secure attachment—the sense of trust in a relationship—and trigger troublesome genes that may activate diseases that might otherwise have remained dormant. The loss of secure attachment is the loss of our best protection against illness—our security blanket.

OUTTAKES

In the United States we have fewer sanctions for parents leaving a hospital with a newborn baby than we do for people choosing to "adopt" a puppy from many animal shelters. In *Ghosts from the Nursery*, we told the story of the Oregon Humane Society that has a policy of requiring potential adopters of puppies to first prepay for puppy shots, worming and neutering. While selected puppies are tagged for prospective "parents," neither the pup not the people can leave the shelter without first meeting with a "counselor" who provides a tutorial in the essentials of puppy health and care and explains puppy needs in depth, including admonitions about the time young puppies require with a person at home (at least half time until the puppy is housebroken!). At the time that *Ghosts* was written, a fenced or contained area outside was mandatory for the puppy, and there was to be a visit by a trained volunteer to the home to assure that these requirements were being met. The American Humane Association, which began as an effort to prevent child abuse and neglect, has found it a much easier task to facilitate the protection of puppies than to provide equivalent safeguards for human babies.

CHAPTER 7

No Place to Hide

The Role of Genetics and Epigenetics

My end is in my beginning.
—T. S. ELIOT

SPURRED BY INCREASING awareness of our own mortality, we baby boomers and our now-adult offspring are looking for answers. What can I do about high blood pressure, weight gain, osteoporosis, heart problems? Does my breast cancer diagnosis mean that my daughter and granddaughter will have to look forward to the same? Does my father's and my grandfather's illness mean that I will suffer the inevitable toll of genetics, or can I avoid my family's health history? Is depression inevitable for me since my mother and grandmother were depressives? Does my parent's addiction mean I am destined for the same? Is it all indelibly slated?

Genetics is fascinating, and this fast-moving science is uncovering answers we all long to comprehend. But the discipline is young and facts are just coming to light, so that what we now know is only the very tip of the iceberg compared to the complex realities we have yet to uncover, particularly those concerning health. At kitchen tables across the country, whether we are searching for answers about a new medical diagnosis or about worrisome behaviors like Dad's alcoholism or an uncle's suicide, an ADD child or a neighbor's violent outburst, we typically ask, "Why?" or "What caused this?" As we wonder about the role of family traits and the influence of traumatic experiences in determining what came about, simple answers can be attractive. "It's in the family" or "The apple doesn't fall far from the tree" are comforting ideas, especially if they help us believe that we are exempt from a given problem by virtue of inherited differences.

A SHORT HISTORY: THE EVOLUTION OF GENETICS

Previously the domain of philosophers, the relatively new science of genetics is now shedding light on how we become and are becoming who we are.

In 1859, the British naturalist Charles Darwin had shocked the world with *On the Origin of Species*, in which he proposed the theory of "natural selection," asserting that traits in a given species evolve over time through interaction with the surrounding environment. The traits that continue to be passed on in any species are those that promote survival and reproduction, while those that do not die out.

Darwin's thinking was augmented by the work of a contemporary, Gregor Mendel, a monk who spent his days in the monastery garden, cross-breeding various types of peas and experimenting with size, texture and color. He theorized that heredity is accomplished through the conveyance of dominant and recessive "factors" to the offspring, half of them bestowed by each parent during reproduction. Preceding both Darwin and Mendel, but enjoying much less acceptance among his scientific peers, was a third pioneer, Jean-Baptiste Lamarck, whose theories were largely dismissed during his lifetime. His "theory of acquired traits," using the unfortunate example of the giraffe, was that changes in the environment can cause the overuse or disuse of an organism's existing structures, which can be handed down genetically to the offspring. According to Lamarck, a lot of reaching by giraffe parents and grandparents resulted in baby giraffes with longer necks. It turns out that he was wrong, but only in the details of the mechanism for evolution.

In spite of his misjudgments about giraffes, Lamarck's observations no longer appear so far-fetched. Ever since Darwin and Mendel, the accepted view has been that traits change only slowly over time and are passed to the offspring through changes in their genes. If, for example, a father develops a "six-pack" abdomen by working out every day of his adult life, his offspring will not, unfortunately, simply inherit his hard-earned belly. Lamarck, who believed, like Darwin, that interactions between species and their environments over a long period of time result in adaptive changes, would be astonished by emerging findings. Genetic research—especially in epigenetics—is casting new light on sacred assumptions about how quickly the environment can alter inherited traits.

CHICKENS AND EGGS

In spite of the fact that science has long since rendered the debate archaic, many still frame questions about the inheritance of physical and behavioral traits in terms of "nature versus nurture." In reality it's impossible to separate the two: nurture shapes and becomes nature, and vice versa. One of the people who says it best—as he so often does—is Dr. Bruce Perry, senior fellow at the Child Trauma Academy in Houston: "Genes are merely chemicals. And without 'experience'—with no context, no micro-environmental signals to guide their activation or deactivation—[they] create nothing. And 'experiences' without a genomic matrix cannot create, regulate or replicate life of any form. The complex process of creating a human being—and humanity—requires both."[1]

More than fifty years ago, the Canadian psychologist Donald Hebb made another memorable comment on nature versus nurture. Asked which he thought contributed more to personality, Hebb responded that asking such a question was like asking what contributes more to a rectangle, its width or its length! Nonetheless, the debate continues.[2]

Most of us are most likely to join the nature-nurture discussion only when we are confronting a new health concern, like cancer, diabetes or depression. In medical and mental health diagnostics, awareness of genetic contributions has tended to prevail. It is now common practice, for example, to view many forms of cancer as strongly genetically linked. And many probably are. But most cancers are seldom a case of "nature" acting alone. If they were, then identical twins who share identical DNA would manifest identical diseases. But they don't; often only one receives a diagnosis. If diseases like hypertension or cancer or schizophrenia are all about DNA, what explains differing susceptibilities between identical twins?

The completion of the Human Genome Project in 2003 provided us with revolutionary scientific knowledge but fell far short of solving several problems that scientists had hoped would be resolved, such as using individual genetic profiles to personalize drug treatment. The reality is that most major diseases are not caused by any single gene or even gene mutation. The major slice of missing information involves complex processes within the gene that surround the genetic material (DNA) and the vast number of potential interactions that can take place between those processes and the external environment. While the sequencing of DNA does not shift in the course of a generation, there is, in fact, another mechanism at work in inherited characteristics. That mechanism is the "epigenome."

Developmental biologist Conrad Waddington first coined the term *epigenetics* in the 1940s; *epi-* means "above," so *epigenome* is meant to convey "above the genome." Epigenetics is the newly emerging branch of biology that deals with the effects of external influences on gene expression. At the biological level, this is where nature and nurture become indistinguishable. The *genome* contains DNA, the blueprints or codes for making the proteins that are the building blocks of life. But DNA is not all that the genome carries. Even more of the genome is made up of noncoding regions that circulate around DNA and regulate how DNA functions, causing certain genes to be expressed while others are repressed. So, along with DNA that remains stable, we inherit dynamic chemical processes within our cells that surround and communicate instructions to our genes, telling them when to be active and when not to be, essentially activating or silencing their expression.

The influences on our DNA include both developmental and environmental factors.[3] If one thinks of the genome as the computer, the epigenome is equivalent to the software that tells the genome how, when and how much to work. This process tells cells what sort of cells they should become—blood, hair, heart. Like a photocopy, each cell in our bodies contains the same genes, the same DNA. It is the epigenome— biochemical activities driven by environmental factors like diet and lifestyle—that silences or facilitates expression of genes in a given cell as they divide to form a given organ. It is this chemistry around DNA that allows cells to differentiate successfully—for example, for stem cells to grow into liver cells rather than heart cells or eye cells as they form a liver.

To confirm some epigenetic realities, take a look in the mirror. You see the reflection of your eyes, your hair, your skin and your teeth, and each of these features looks very different from the others. But the DNA in each feature is identical. How is it that cells that form our heart, eyes and liver are very different cells? The proteins that accomplish epigenetic processes function like microscopic satellites that rapidly respond to their surroundings. A nerve cell doesn't need to use genes that a liver cell might require, just as a liver cell doesn't need the genes that a heart cell requires. It is epigenetic chemistry that causes cells to pack away unneeded DNA, either by tagging such extra baggage with a small chemical or by wrapping it into a dormant form. Viral infections, hormones, some aspects of diet and exposure to toxins like alcohol, prescribed and nonprescribed drugs, or environmental pollutants can cause the epigenome to go into action, altering gene expression. Scientists such as Vardhman Rakyan of the Sanger Institute believe that up to 70 percent of the contribution to a given disease is "nongenetic"—that is, epigenetic.[4] As if this wasn't complicated enough, in the course of one lifetime epigenetic messages may be passed from parents to offspring through chemical "tags" that have been added onto parents' DNA. Enter the fertile field of epigenetics—the study of the interactions of one's genes with one's environment. This is the new frontier in research. And it is revolutionizing our world.

We have traditionally thought of genes as mechanized programming by which a fertilized egg grows automatically into an adult. In fact, this is far from a closed process. Genes are in fact amazingly plastic, so that an organism can take in information from its surroundings and adopt a survival strategy. The result is a unique developmental course that affects growth patterns, chances of survival and reproductive success. This capacity for adaptation is at the core of the epigenetic structure. But what is initially an asset can become a liability, as we will soon see when we look at the roots of disease. Let's look first at adaptation as an asset for survival.

The alpine newt, a primarily aquatic creature that lives in the lakes of Switzerland and France, is a good example. The baby newt, looking quite like a tadpole, hatches from eggs laid in the shallows at the edges of the lakes. The newts' epigenetic adaptations make available two very distinct courses for adult development. Some remain as tadpoles, retaining their infantile appearance. Those that stay infantile can swim deeper into the lakes, where they have more plankton to eat and less competition, so they grow faster and have more reproductive competence. In other newts, typically the larger infants, the gill slits close over so that they can live on either land or water, enabling the species to have a greater likelihood of survival when water levels fall. By assuring a flexible adaptation of a basic blueprint, the species is more likely to survive.[5]

Unfortunately, the same sensitivity that allows communication between a given cell and surrounding cells, and between all the cells in the organism and the organism's environment, can also facilitate destructive changes. All that we eat, drink, smoke and feel influences this process. A mechanism within a cell that is perfectly designed to enable us to adapt can thus become the catalyst of our demise. The same exquisitely tuned mechanisms that trigger some cells to grow to form eyes while others become skin can also cause the growth of cancer.

To understand this further we return to the studies of identical twins. Imagine for a moment being one of two people who are genetically identical. Emerging from the same egg, you and your twin have precisely the same genes. You are essentially clones of each other. Appearance, mannerisms, expressions, senses of humor, and what you choose to wear on a given day (even without knowing your twin's choices) are all identical. Most people who know both of you perceive subtle differences between you the older you get. These dissimilarities can actually be confirmed by geneticists. With the help of a powerful microscope, they can view the increasing differences between your cells and your twin's as you age, even though, as identical twins, you started out the same in the beginning, with precisely the same genes, and the sequence of your identical DNA has not changed.

Now imagine that in adulthood your twin is diagnosed with cancer. How did this happen? What does this mean for your own health? Historically, medical science has believed that it was only a matter of time until you would suffer the same fate. But this may not in fact happen. What is going on?

A GHOST IN YOUR GENES?

For answers to questions about identical twins and the prognosis for disease when one has already been diagnosed, geneticist Dr. Randy Jirtle of Duke University has some surprising answers.[6] Jirtle studies genetic differences by looking at "agouti mice," a breed named for a particular gene that they carry: the agouti gene. A first look at a pair of agouti mice in Jirtle's laboratory reveals two distinctly different-looking types of mice. One mouse is golden in color and fat, so fat that it cannot move easily. The fat yellow agouti is constantly ravenous and prone to cancer and diabetes. The other mouse is brown and skinny, weighing about half what the yellow mouse weighs, but like identical twins, the mice are genetically identical. Dr. Jirtle explains that while the mice share identical DNA, in the fat yellow mice the agouti gene stays permanently "on," causing obesity, while in their skinny brown siblings the agouti gene is "silenced" or shut down by epigenetic processes.[7]

One way to silence a gene is to prevent the molecular machinery from accessing it and shutting down its ability to influence protein formation by wrapping it in an inaccessible package. In the case of the agouti mice, as in the case of identical twins where one has a disease but the other does not, scientists can detect a tiny chemical tag of carbon and hydrogen, called a *methyl group*, at work on the gene. In the brown mouse, the tag attaches to the agouti gene to inhibit or silence its programmed function. Most living creatures have millions of these tags attached directly to their genes, inhibiting their function. DNA expression is thus chemically altered in spite of the fact that its sequence is untouched. This process is called *methylation*.

A second type of epigenetic process involves *histones*, the protein spools around which DNA tightly coils. At the microscopic level, DNA is arranged something like a tight slinky toy. It is this packaging that allows the considerable length of DNA molecules to fit inside a cell. In its tightly packed state, DNA cannot be used as a protein template. To become usable, the DNA has to be exposed and unraveled. In a process known as *histone acetylation*, the attachment of another type of molecule, acetyl molecules, to part of the histone causes the packaged structure to expand and unravel a bit, enabling gene expression.

These two processes—methylation and histone acetylation—through differing chemical responses turn genes in the cell "off" or "on." The genes that a given cell does not use or need are turned off or "silenced" by methylation, while portions of DNA necessary for that cell to grow are expressed through histone acetylation. This communication of instructions that influence DNA activity without altering DNA sequence is epigenetics at work.

Some scientists now view this distinct pattern as a kind of second genome—the "epigenome." Epigenetic changes are by design sensitive to environmental influences; this is how toxic chemicals and lifestyle affect our susceptibility to disease. Still in the trial stages of medical research, epigenetic therapies target the alteration of methylation and histone acetylation patterns to allow expression or suppression of genetic messages. For example, in cancer therapies, epigenetic therapy can silence the expression of cancer-causing genes or stimulate tumor-suppressing genes.

In the case of fat yellow and thin brown agouti mice twins, a pregnancy diet rich in methyl donors like folic acid and vitamin B-12 makes huge differences in both the appearance and the health of the offspring without altering a single letter in the DNA. The dietary change forms tags that turn off the agouti gene, changing the color of the coat and the tendency to weight gain along with the predisposition to disease. When the coats of baby mice are brown, scientists know that they will be the skinny and healthier variety. Timing is important, as it is with humans. Beginning this process during critical periods such as pregnancy, perinatal development or puberty can make crucial differences. The take-away message here, researchers warn, is not simply that folic acid is good for pregnant humans, though some is essential. The most important point to understand is that environmental factors like nutrition can have a dramatic impact on our inheritance of health and disease. And it may not end with one generation. The epigenetic change in the agouti mice—the silenced gene marked by the brown coat—was inherited by the grandpups of the fat mothers, regardless of what their mothers ate. Moreover, when an environmental toxin was added to the grandmother mouse's diet, such as biphenol A (BPA), a toxic ingredient used in the manufacture of plastic containers like water bottles, most of the grandpups were born with yellow coats and the tendency to be obese and diabetic.

Until recently it has been heresy in scientific circles to assume that people's experiences might alter the traits inherited by their descendants without changing a gene sequence. For example, no one expects that a mother's shapely nose attained through plastic surgery will produce similar assets in her offspring. Changes in physique acquired by parents will not be reflected in their children because these changes don't affect their genes. Yet epigenetic findings have us looking seriously at the effect of parents' diet, stress and other factors on the next generation and beyond. Although we have long known about the impact of toxins like BPA ingested during pregnancy,[8] we now know that the effects may be passed on to grandchildren, great-grandchildren and beyond. We have long believed that alterations in the expression of parental DNA by methylation or histone activity are erased in eggs and sperm, that the slate is wiped clean for the next generation. The agouti mouse is just one example showing that this is often not the case.

In 1999 the Australian biologist Emma Whitelaw, of the Queensland Institute of Medical Research, proved that epigenetic marks can be passed from one generation of mammals to the next. "It changes the way we think about information transfer across generations," Whitelaw says. "The mind-set at the moment is that the information we inherit from our parents is in the form of DNA. Our experiment demonstrates that it's more than just DNA you inherit. In a sense that's obvious, because what we inherit from our parents are chromosomes, and chromosomes are only 50 percent DNA. The other 50 percent is made up of protein molecules, and these proteins carry the epigenetic marks and information."[9] With a nod to the long-defamed Lamarck, recent years have produced numerous reports on these new findings, often featuring humorous titles like "The Sins of the Fathers" and "I Blame My Grandparents."[10]

Nature is full of examples of this generational legacy. A fascinating one is the tiny water flea, which is not simply a wet version of the insect that causes people and dogs to itch. Rather, it is a tiny opaque member of the crustacean family ranging in size from microscopic to more than two inches. The water flea's inapt name comes from the jumping or hopping motion it makes as it moves through the water; they dine on algae and plankton, not people. Water fleas are translucent, so one can easily see their eyes, antennae and internal organs at work. But what is most interesting about them is what can't be seen: although they share identical DNA, some have a spiny helmet on their heads, while others are bare-headed. Why are some protected while others remain exposed? The answer lies in the experience of the mother water flea. If she had a bout with a predator, typically a bite, her babies are born with helmets. If her life was not threatened, her babies are bare-headed. Along with, but not affecting, her DNA, the experience of the mother is transmitted to her offspring.[11]

Mothers are not alone in their epigenetic influence. Professor Marcus Pembry, a clinical geneticist at the London Institute of Child Health, adds fathers and grandfathers to the list. His study published by the University of Bristol in collaboration with Umea University in Sweden has potentially far-reaching implications for health epidemics, including obesity and cardiovascular disease. Together with his partner Dr. Lars Olor Bygren, a preventive health specialist at the Karolinska Institute in Stockholm, Pembry combed through records from a remote parish in nineteenth-century Sweden, noting patterns of harvest and food prices, periods of food scarcity or bounty, and the health of three generations of families. He found a puzzling pattern: the life span of grandchildren appeared to be influenced by their paternal grandfathers' access to food, especially during the grandfathers' pre-puberty growth period, from nine to twelve years of age. If the grandfathers had endured food scarcity at this particular stage, their grandsons were less likely to die of cardiovascular disease than the grandchildren of men who were not exposed to famine during this developmental stage. Similarly, having a grandfather who had lived through famine was also associated with a lower risk of diabetes. In fact, if the grandfather had plenty of food during this period, his grandson was about four times more likely to die with diabetes! This was one of the first indications in the research of the transgenerational effects of experience.

Digging deeper into the unusually detailed records, the researchers looked at the food supply every year for both grandfathers and grandmothers, from their conception to age twenty. They found a similar effect between grandmothers and granddaughters. But in the case of the females, the sensitive period for the grandmother's exposure occurred when they were yet in the womb. Clearly there were sensitive periods for both. But why the difference? Why would pre-puberty be key for the men while gestation was the critical correlate for the women? What was the common denominator? The researchers determined that gestation and puberty are critical to the development of the eggs and sperm and that the environmental information—a kind of physical "memory"—was being imprinted on the egg and sperm at the time of their formation. The granddaughters exposed to famine in the womb, when girls form all of the eggs they will ever have, were more likely to die sooner than their peers whose grandmothers had suffered no such exposure.

Researchers are still unable to answer the question of why the data seem only to apply to same-sex progenitors. That is, why did grandmothers' experiences only affect granddaughters and grandfathers' only grandsons? Pembry and others wonder whether the current epidemic of obesity might be tied to the lifestyles of our forebears, two or three or more generations ago. Regardless, the idea that famine can be captured by genes without affecting their sequence and be carried forward to future generations is giving scientists a whole new take on the meaning of inheritance.[12]

The "Dutch Hunger Winter" is another example of the generational transmission of experience. In the winter of 1944, the Nazis banned the use of railroads to transport food to western Holland in retaliation for a transportation strike staged by Dutch citizens to demonstrate their opposition. While shipping continued on the canals for a while, they froze during what was a particularly harsh winter, seriously aggravating the shortage of food. The resulting famine lasted for several months until the Allies liberated Holland. Throughout the famine some Dutch hospitals kept detailed medical records and then followed the health of the affected population and their children over several generations. They found that girls who were malnourished in utero during the first trimester of pregnancy were born at normal size, but often, as adults, gave birth to smaller than normal babies. Subsequent research has revealed that girls born small for their size typically have smaller uteruses, which in turn constrain the growth of the next generation of babies.[13]

These findings echo the work of David Barker, who believes that health in midlife is principally determined during fetal life and the first two years after birth. Barker compares the offspring of the Dutch mothers during the famine to the children of Helsinki born between 1933 and 1944 whose birth weights were below six and a half pounds and who were thin for the first two years of their lives. These babies became the adults most prone to heart disease as adults.

As the Dutch continue to monitor the health of the now middle-aged offspring of the 1945 famine, they are seeing a much higher rate of chronic diseases, including diabetes, kidney disease, and almost triple the rate of heart disease, than is seen in individuals born before or after the famine. Twice as many in this cohort rate their health as poor compared with those born before or after the famine. When these data came to light, several researchers were skeptical of Barker's explanation. One critic, Dr. Douglas V. Almond, an economist at Columbia University, initiated a more rigorous study to determine whether other factors had contributed to the Dutch famine outcomes. Looking for an event that would have affected everyone in a given population—rich, poor, educated or not—and then just as suddenly dissipated, he settled on the 1918 flu pandemic. Almond compared the Dutch Hunger Winter population with the flu epidemic that hit the United States between October 1918 and January 1919, afflicting one-third of all pregnant women in the nation. Comparing the offspring of the affected mothers with those who had been born just before or just after the pandemic, Almond found the same pattern as in the Dutch subjects. The adult offspring of women who had been pregnant during the epidemic had more illness, especially diabetes, which was 20 percent higher by the age of sixty-one. They attained less education, had lower incomes, and were more likely to be on welfare. The effects were seen across race, across incomes, and in both men and women. Saying the data spoke for itself, Almond endorsed Barker's conclusion that what happens before age two permanently affects our health, including the aging process.[14]

It is increasingly evident that, along with negative experiences such as famine and stress, environmental toxins take their greatest toll during earliest development. Echoing the work of Randy Jirtle and his agouti mice, Dr. Matthew Anway, a researcher with Washington State University, has found that exposing rats to noxious doses of a fungicide in utero can induce disease in subsequent generations of rats. This conclusion,

that contaminant disease states might be inherited by multiple generations through epigenetic mechanisms, has huge implications for human
diseases such as Parkinson's, amyotrophic lateral sclerosis (ALS, Lou
Gehrig's disease) and many forms of cancer, including breast and
prostate. Though the applications to the human population have not yet
been researched, this line of inquiry may reveal a blind spot in the etiology of several chronic adult diseases.[15]

Toxic exposures are increasingly implicated in generational effects. In
2005 Anway reported that in rodents a toxic developmental exposure could
decrease sperm count not only in the exposed offspring but in at least three
subsequent generations. "In human terms," Anway notes, "this would
mean that an exposure experienced by your great-great-grandmother while
your great-grandfather was in her womb could have an adverse effect on
your sperm count!"[16]

SUBTLE DISRUPTIONS IN NORMAL DEVELOPMENT

Interference in the earliest reproductive stages is increasingly implicated
in several diseases and may sometimes be an unseen outcome of in vitro
fertilization. Both men and women have twenty-two pairs of *autosomes*
(chromosomes that are not sex chromosomes) and one pair of sex chromosomes. The female sex chromosome is always XX, and the male is always XY. During the creation of the eggs and the sperm, the pairs of sex
chromosomes of each, the male and female, divide. Each of the female's
eggs carries one X chromosome, while each sperm carries either an X or
a Y chromosome, so that the sperm determines the sex of the offspring.
All of the other chromosomes also divide so that the baby gets half from
each parent. But in a small subset of chromosomes, one copy of either the
mother's or father's genes can be "imprinted"—or turned off in either
the egg or the sperm.

There is an unseen war at work between the directive from the father's
genes and that of the mother's. Imprinted genes from the father push to
promote growth, while the mother's genes tend to suppress growth. The
maternal genome seeks to limit the extraction of nutrients from the
mother while the paternal genome works to enhance it. Imprinted
genes can trigger various pathologies because they are easily dysregulated,

especially if they are tampered with at this critical stage—as they can be during in vitro fertilization (the process of trying to replicate the complex female reproductive system in a petri dish). As careful as these procedures may be, imprinting errors are relatively more likely to occur and often manifest as developmental and neurological disorders when they occur in early development or as cancer when alterations manifest later in life.[17] In vitro imprinting disorders have been linked to Prader-Willi syndrome, Beckwith-Wiedemann syndrome, Angelman syndrome, Alzheimer's, autism, bipolar disorder, diabetes, male sexual orientation, obesity and schizophrenia, as well as most forms of cancer.[18]

Mismatch

There are strong indications that we are not only the victims of environmental challenges, such as episodic starvation, but that we are now also the perpetrators of such challenges as we inadvertently create an environment increasingly out of tune with our bodies. Having evolved in small groups or clans, essentially living as hunter-gatherers for most of our time on the planet, our current cultures and lifestyles are far removed from the daily lives that shaped our biology.

Scientists like Peter Gluckman and Mark Hanson believe that this mismatch between our bodies and our current lifestyle is responsible for the deadly explosion in lifestyle diseases such as heart disease, diabetes and obesity. We are confronted daily by evidence that we must do something differently. Understanding what that is and how to effect change is the topic of their powerful book, *Mismatch: Why Our World No Longer Fits Our Bodies.* They warn of the growing dangers that result from the products and by-products of technological manipulations of ourselves and our environment, citing as examples the falling age of puberty and its unfortunate conjunction with the rising age of psychological maturity and our increasing longevity, now accompanied by rising rates of Alzheimer's and dementia. What can we do? Drawing upon epigenetic research, Gluckman and Hanson direct us to the beginning of the life cycle, saying that in order to understand disease we must understand human evolution and development. They recommend interventions targeted early and doing everything possible to scaffold the health of potential mothers.

Mothering Matters

Factors beyond foods and chemicals—especially maternal attention—can have an impact on gene expression. One example is Dr. Michael Meaney's "high-licking-and-grooming" mother rats that produced pups that were stress-resilient throughout their lives. Like nutrition, a mother's attention imparts long-lasting changes in the genes that affect behavior. In 2004, Moshe Szyf of McGill University in Montreal demonstrated that rats that don't receive extra attentive behavior from their mothers develop a methyl tag on a gene vital to the modulation of the stress response, shutting off this potential. Szyf and his colleagues are now considering how this applies to humans.[19]

TWO TO TANGO:
SEARCHING FOR THE ROOTS OF ILLNESS

While many diseases, including several forms of cancer, result from the interaction of our physiology with negative factors in our environments, the converse is also true: health and healing are imparted from constructive environments, particularly relational experiences in early life. Dr. W. Thomas Boyce, a principal investigator with the Canadian Population Health Initiative, refers to this interaction of individual biology with environmental factors as "symphonic" causes of disease and health. The theory is that disease results from the conjunction of individual susceptibility with external threat, and that health results from the match between one's needs and vulnerabilities and what the environment provides. This strand of research looks closely at "reactive" individuals, those who are the most susceptible to disease and morbidity. Rather than being a genetic rarity, the individuals who are at highest risk of illness and death from the interaction of biological sensitivities with environmental risks are typically 15 to 20 percent of any given population. Such sensitivities appear to be cumulative, often beginning with a highly stressed gestation, infancy and childhood, gradually building on themselves, and ultimately catalyzing disease processes like inflammation and immune dysfunctions.[20]

New research on the epigenetics of major diseases is moving so rapidly that it is almost impossible to remain current on the findings. Given the complexities, we limit our discussion here to a sampling of the diseases most highly correlated with stress and trauma early in life and implicated in the ACE Study.

EPIGENETICS AND PHYSICAL DISEASE

Although many, if not most, forms of cancer are not necessarily linked to early development, cancer was the first disease that researchers examined through the epigenetic lens, and what they have found so far has implications for other diseases. Remember that all of our genes are present in all of our cells. All of our tissues—hair, skin, heart, eyes—are genetically identical. But at any given point, a tissue may utilize only 10 or 20 percent of the genetic material available. The rest is packed away so that these genes can't shape formative proteins in a given cell. Epigenetic processes silence unneeded genes, while genes that are necessary for the job of a given organ are protected from silencing. Until recently, cancer has been viewed as resulting from gene mutations—gene structures that become abnormal. Scientists now believe that more forms of cancer result from faulty instructions from the epigenomes to silence or express genes than from inherited DNA mutations. Different cancers can result from either or both, which is why different cancers respond to different types of therapy. Known risks for cancer—for example, cigarette smoking or sun or radiation—can trigger the disease either by direct genetic damage or by triggering an underlying epigenetic process: the abnormal aging of cells.

Cells within our tissues normally live only for weeks or months. Stem cells divide to replenish tissues when they are damaged. When cells have to divide hundreds of times, there will be subtle errors, tiny epigenetic shifts that accumulate with age. Every time a stem cell repairs an injury it ages a bit more, so that aging is actually a reflection of how many times stem cells have had to divide. Some cancers are, in effect, old tissues that contain cells that may have divided hundreds of times depending on how long the cancer has been active. So the common denominators in many forms of cancer are tissue damage, inflammation and the need for stem cells to divide to repair the injury. The more injury and repair, the older and more at risk of cancer our tissues are.

Cancer

Epigenetics promises exciting frontiers for cancer therapies by reversing epigenetic aberrations. Although still in the early stages, research by Dr. Alfonso Dueñas González at the University of Mexico has begun, with some very promising results, to explore gene reactivation through a combination of drugs designed to use both methylation and histone inhibition. The combined drugs reactivate the genes involved in cell proliferation and immune recognition of tumor cells, along with many other processes involved in cancer detection and defeat. Unfortunately, the drugs being tested in Dr. Dueñas González's lab, having been used for over thirty years to treat nonmalignant conditions such as hypertension, are no longer eligible for patent protection and thus are less interesting to pharmaceutical companies, which are major sources of funding for such research.

Though still in its earliest stages, Dueñas González's epigenetic therapy for some forms of cancer—for example, myelodysplastic syndrome (MDS), often known as "pre-leukemia"—is the new frontier for treatment. When doctors look at the bone marrow of patients with MDS, they see 99 percent cancer cells that are copying themselves tirelessly, crowding out normal tissue and preventing normal function. All of the cells that accomplish what bone marrow cells normally do—carrying oxygen, making red blood and white blood cells and platelets—become abnormal and greatly reduced in number as MDS progresses. Historically this has been a death sentence as patients succumb to bleeding, anemia, heart attacks or infections. The goal of epigenetic therapies, as opposed to most forms of cancer therapy, is not to kill the cells but to change the instructions to the proteins surrounding DNA within the cells and to reactivate genes that have been silenced. These drugs work by reversing methylation—that is, adding or removing methyl tags from certain cells. The initial results of this epigenetic therapy—relatively low doses of the drug Decitabine—for more than one hundred MDS patients have been remarkable: with minimal side effects, many have had complete remissions; almost half have seen the complete disappearance of the disease. A small proportion of patients remain unaffected. Researchers believe that the results are easier to attain with MDS than with many other forms of cancer because MDS is a blood cancer that does not involve a tumorous mass; the next horizon will be epigenetic therapy for tumor-causing cancers, including those affecting breast, lung and skin.

Because of the dynamic nature of the epigene, there are huge implications for modification—especially in earliest development—of some diseases through diet and nurture as well as through exercise and lifestyle choices, including nicotine and sun exposures. We are born with our genome, but we can protect our epigenome.[21]

We used to think that our genetic code was laid down from the beginning of life and that whatever we got was a life sentence. But the new understanding of the epigenetic code brings great optimism to this often-bleak prognosis. Epigenetic patterns can change through life, and little changes can create huge differences. Researchers like Moshe Szyf are showing us how. Szyf, a pharmacologist, has found that the DNA methylation machinery is key: knocking out one of the enzymes in the pathway that leads to cancerous tumors can prevent tumors from developing. Adding or removing methyl tags can increase or suppress gene-directed tumor growth. Too little or too much methylation can activate cancer-causing genes or silence tumor-suppressing genes. The promise is that epigenetic changes are potentially reversible, unlike genetic mutations. While the only recourse with a mutated gene is to kill the gene, genes with a defective methylation pattern can be reset to a healthy pattern.

There are several spin-offs of the epigenetic research. One is testing for inherited epigenetic effects, like the humans who weathered the Dutch Hunger Winter. Medical research companies are developing tests for the heritability of colon cancer as well as breast and prostate cancer in the hope that early identification might lead to prevention. Other researchers are investigating the role of diet in protecting human epigenomes. Some geneticists, like Ignatia van den Veyver of Baylor College of Medicine, believe that "methylation diets," tailored to individual profiles, may be the key to staying healthy in the face of various toxins. Certain foods such as green tea and soy are being studied for their protective properties in preventing methylation from shutting down certain cancer-fighting genes. Researchers across the world are hoping that the new international Human Epigenome Project will advance this promising frontier by centralizing data and setting standards for the technology essential to decoding epigenetic patterns.[22]

We have all heard of individuals who experience remissions of cancer or who live well beyond their projected life expectancy subsequent to being given a death sentence with a terminal cancer diagnosis. It is quite likely that the epigenome is at the center of these stories. Epigenomic changes—small changes in how we live, including how we feel, how we relate, and with whom—can make big differences for many of us throughout our lives, even under end-of-life conditions.

Diabetes

While the most prominent target of epigenetic therapies has been cancer, there is also great interest in the epigenetic mechanisms at work in type 2 diabetes, a disease that has now become epidemic. Researchers have discovered a connection between intrauterine growth retardation and the development of type 2 diabetes later in life. Studies on rodents are focusing on protein and zinc deficiencies in the prenatal environment that may trigger epigenetic adaptations that cause or contribute to the disease.[23] Pembry and Bygren's study of the nineteenth-century Swedish famine indicates that epigenetic changes due to dietary patterns may be handed down over generations. People whose grandfathers overate as children evidenced a higher incidence of diabetes, while those whose grandfathers ate less had a lower rate of heart disease. Grandpa's overeating at the smorgasbord as a lad probably influenced how his diabetes-related genes were methylated—and how active they were. His children and grandchildren inherited not only his DNA but also his methylation pattern. While Grandpa's eating habits may have been matched to his lifestyle, they do not fit today's more sedentary one. So now his grandchildren may be overweight and prone to diabetes. The implication is that methylation imbues cells with a sort of memory of the experiences of one's ancestors.

As in Barker's hypothesis regarding future heart health, diabetes now appears to be a disease based on the mis-preparation of the organism for the environment. Dr. Rama Natarajan, a professor in the Department of Diabetes, Endocrinology and Metabolism at City of Hope in California, has been investigating how this occurs in diabetes by studying inflammatory and immune cells grown in high-glucose conditions, similar to those affected by diabetes. She found epigenetic changes in several genes known to be related to diabetes, a first step in a long process of hunting down ways to intervene in this disease.[24] There is a strong relationship between type 2 diabetes and obesity, which in turn correlates with several other diseases, including hypertension, stroke, kidney disease, breast cancer, Alzheimer's and the biggest killer of them all—heart disease. Several key diseases share a few underlying processes, like inflammation. So a major discovery in any one of these diseases may catalyze a domino effect in therapies that could turn these processes "off" or "on" to stem such conditions.

Heart Disease

It is well known that heart disease is the leading cause of death for adults in the United States. What is less well known is that it is also the leading noninfectious cause of death for children under age one. Unlike cancer, which was linked to epigenetic causes in the early 1980s, heart disease has not yet yielded its mysteries to epigenetic research. We know that there are environmental risk factors for heart disease, like smoking and diet, in addition to genetic ones. But understanding how environmental and genetic factors interact to create heart disease is still in the earliest stages of research.

In summarizing the research at Japan's Jichi Medical University on the connections between epigenetic changes and heart disease, a spokesman said: "Recent genetic and biochemical analyses indicate that epigenetic changes play an important role in the development of cardiac hypertrophy and heart failure. . . . The results of these studies should not only improve our understanding of the molecular basis for cardiac hypertrophy/heart failure but also provide essential information that will facilitate the development of new epigenetics-based therapies."[25] Another Japanese researcher, Kunio Shiota, a professor of cellular biochemistry at the University of Tokyo, says simply: "The bottom of the iceberg is epigenetics."[26] There is little question that stress, trauma, diet and exercise play a role in heart disease, or that the effect is cumulative. Early stress and trauma are stealth agents with epigenetic impact. But science is still far from producing non-surgical remedies that lessen the toll once heart disease is full-blown.

In taking a closer look at epigenetic changes resulting from early chronic stress, a team of researchers at McGill University in Montreal has found that adverse childhood experiences, such as child abuse, leave epigenetic marks on DNA that mediate the stress response, resulting in higher rates of anxiety, depression, substance abuse and suicide. They compared the brains of child abuse survivors who committed suicide as adults with those of suicide victims who were not abused as children. In the brains of people who had been abused, the genes responsible for clearing cortisol were 40 percent less active, rendering them significantly more vulnerable to stress. This research documents that how children are treated registers at the cellular level and provides a missing link in the correlations found in the ACE Study.[27]

Asthma

Although asthma is not one of the major diseases implicated in the ACE Study, it is included here because it emerges in the epigenetic research as one with suspected links to earliest experience. Asthma is typically thought of as a genetically conveyed allergic response that is exacerbated by environmental factors such as cigarette smoke and allergens. But a team of researchers from Mayo Clinic along with researchers from two Colorado facilities report that children genetically susceptible to asthma have more than double the risk of developing the disease by the age of eight if their parents had trouble caring for them in the first year of life. The study followed 150 primarily Caucasian children. The researchers speculate that the linchpin is emotional stress to the baby that impairs development of the immune system. Both marital conflict and maternal depression were identified as factors that catalyzed stress to the children. The emotional caregiving environment, not larger environmental factors such as low socioeconomic status, was shown to have an impact. Conversely, sensitive and responsive caregiving appeared to have a buffering effect, gentling the reactivity of the HPA system.[28]

A DISCLAIMER

I can never reflect on the impact of early experience on genetics and epigenetics without a disclaimer. I have dear friends who carry a dreadful inherited disease called Fanconi's anemia, which is caused by a rare genetic

mutation that both of these parents carry. They are brilliant professionals
and have had four children, three girls and a boy. All three girls have had
the misfortune of being born with Fanconi's anemia. The disease has already taken the lives of the two eldest daughters and now, in the absence of
a medical miracle, threatens the life of the third. Clearly there are genetic
diseases that proceed on course with little epigenetic modification from
even the most optimal environment, although it is my belief that the girls,
each of whom was amazingly accomplished, lived as long as they did—
through late adolescence—because of the quality of their early lives.

It is obvious that epigenetic influences do not trump genetics for those
who are living with a genetic disease, such as multiple sclerosis (MS),
amyotrophic lateral sclerosis (ALS), epilepsy or type 1 diabetes. Epigenetics is still only part of the equation, and the research is far from offering simple cause-and-effect answers. Nevertheless, the quality of
human connections that surround us, especially when we are most vulnerable, are clearly life-enhancing.

EMOTIONAL AND BEHAVIORAL HEALTH:
ALL IN THE FAMILY?

Researchers have spent years trying to locate a "schizophrenia gene" or an
"alcoholism gene." There have also been attempts to isolate the genetic
basis of "resiliency." All such efforts have failed and yielded no clear answers beyond the finding that multiple genes are indicated in each disease
and resilience. Contrary to popular thinking, neither the tendency toward addiction nor toward mental illness nor toward resiliency in the
face of misfortune is attributable to genetics per se. Most researchers see
these tendencies as resulting from the interplay between specific genes
and the environment, or "G × E."

Genes are simply potentials—like building blocks. When it comes to
emotional and behavioral pathologies, vulnerable gene variations can do
no more than promote a given tendency, which may never surface in the
absence of triggering environmental factors. So, for example, genetic variations promote resilience in those who have the genes associated with resiliency. But, according to Sir Michael Rutter, a British pioneer in child
psychology, these genes in an individual never exposed to childhood

trauma "don't do anything much on their own." Similarly, an individual with two addictive parents, in the absence of early chronic stress or trauma, may never express the genetic proclivity for addiction, or may express it only minimally. The confounding issue, of course, is that in a family with two addictive parents, trauma is very likely. In most major mental illnesses, several genes are implicated, but in the absence of early trauma, these illnesses may never appear.

Addiction

According to the National Institute on Drug Abuse, drug addiction is a chronic disease characterized by changes in the brain that result in a compulsive desire to use a drug. Approximately 10 percent of all people who experiment with drugs become addicted.[29] In the ACE Study, men with an ACE score of 6 or more were 4,600 percent more likely to use intravenous (IV) drugs.

Researchers studying the genes connected with addiction begin by comparing the DNA sequences of addicted and nonaddicted individuals. The next step is to narrow down the possibilities and identify a small number of so-called candidate genes. Typically beginning with mice, a suspected gene is identified. For example, a gene that appears to affect the reward pathways is identified in rodents who are drawn to alcohol consumption. By comparing DNA sequences (which are remarkably alike in mice and men!) the corresponding gene in humans is then identified. Genes influence addiction by affecting pathways in the body and the brain—processes that interact with each other and with the person's experiences—to promote vulnerability or protection. Technology now allows researchers to survey hundreds of thousands of tiny variations in the genomes of each individual, pinpointing influences on his or her physiology and individual risk for disease.[30]

Some substance-specific genes are implicated in addiction—for example, gene variations that influence the inability to metabolize alcohol. Other genes pertinent to addiction are related to mood, anxiety and personality disorders. Still others appear to influence brain pathways involved in reward-seeking, by triggering the brain's receptor cells for serotonin and dopamine, and there are genes associated with taking risks

and seeking novel experiences. Each of these types of genes is implicated as a candidate gene, meaning that it may play a role in either the
risk of addiction or protection from addiction. Dr. David Goldman of
the National Institute on Alcohol Abuse and Alcoholism says: "The
stronger the drug, the stronger the role of heredity in causing an addiction to it." If this is so, genetics may have more to do with some people
becoming crackheads than potheads, who may be more influenced by
their social group.[31]

Whether we are attracted to champagne or cocaine, whether we are
sickened or easily hooked by narcotics—these responses are under some
degree of genetic control, as are differences in how we handle the effects
of alcohol (getting dizzy, nauseated, sleepy or easily drunk) and our risk
of addiction. Those who are least sensitive to unpleasant side effects are
the most prone to addiction. For people who carry genes that cause them
to flush or develop a rapid heartbeat and nausea when they drink, alcohol can be a carcinogen leading to esophageal cancer. More than half a
billion people carry these genes, including approximately 36 percent of
Chinese, Japanese and Koreans. None of the candidate genes exert an influence across the board for all people, and several of the candidate genes
for addiction are also recognized as playing a role in some forms of mental illness, such as depression and anxiety, and learning differences, such
as ADHD, which are often comorbid with addiction.[32]

Still, we can't blame genes alone for addiction. As in heart disease,
obesity and other diseases, "genes merely load the gun," researchers
note, "while environment pulls the trigger."[33] Trauma, abuse and neglect in childhood are known to be associated with environments that
contribute to drug abuse and addictions of all kinds, but some people
nevertheless manage to avoid addiction. One possible explanation may
be variants in the gene that controls an enzyme that helps regulate the
brain's stress response. Those of us who carry a variant of this gene and
were also abused as children are much more likely to become alcoholic,
but without early abuse, the gene variant alone has no effect on alcoholic tendencies.[34]

On an individual level, susceptibility to addiction or to mental illness
remains hard to identify genetically. Even when genes can be identified
that are associated with aspects of addiction, genetic susceptibility does

not mean genetic destiny. It may mean it is more difficult to quit once a person starts, or that withdrawal will be more severe, but no one is doomed to become an addict. Various genes and environmental factors can either accumulate or cancel each other out. To make things even more complicated, risky genes may vary from one person to another, and not everyone who carries the risky genes will exhibit addictive tendencies. Conversely, people without the genetic proclivities may also become addicted in spite of the fact that they have a genetic tendency to feel sick from a drug that would make an addict feel good. There are many routes to addiction for any of us.

There are several suspected pathways involved in alcohol addiction, some of which also have implications for addiction to other substances. One involves the breakdown of alcohol itself. Although no one ethnicity is more prone to alcoholism than others, certain of the genetic variants that contribute to risk are much more prevalent in some ethnic groups than in others. An example is the ADH4 gene, which creates a slow-working enzyme that makes many Asians intolerant of alcohol, causing potentially toxic side effects from drinking. Another genetically linked variable is the degree to which substances interfere with the balancing of brain functions such as impulsivity versus inhibition. Of particular interest is gamma-aminobutyric acid (GABA), the most common neurotransmitter in the nervous system; it modulates impulsive behavior by inhibiting a cell's responsiveness to signaling. The gene variations that govern GABA production and the degree to which substances interact with this function influence one's ability to modulate impulsive behaviors, a key factor not only in alcohol dependence but also in bipolar illness and conduct problems. Variations in another gene, CHRM2, which is linked to cognitive functions like decision-making and attention, affect the balance between limbic-driven impulses and cortical restraints and are linked to depression and other internalizing symptoms that, in turn, foster drinking. The impact of such genetic variations sounds suspiciously similar to the known effects of childhood trauma on the brain and is probably one of the underlying factors at work in the ACE correlations. These gene variants affect the functions of the frontal cortex, a part of the brain that is under construction during the first years and is programmed for vulnerability or resilience by early nurturing.

Improved knowledge can lead to better decisions. Historically, when Americans began to comprehend the risks of nicotine, as well as secondary smoke, lung cancer, asthma and other lung diseases, smoking declined. As the research moves forward, genetic profiling is expected to provide individualized medical, behavioral and emotional information and recommendations that could make a huge impact on our species over time.[35]

In the burgeoning field of addiction treatment, the new wave of thinking points increasingly to the effects of epigenetic responses to various and accumulating forms of trauma, especially early trauma. Specialists in trauma, like Robert Scaer, and attachment experts, like Allan Schore and Daniel Siegel, are headlining the training forums for addiction professionals, and the protocols for dual-diagnosis treatment are shifting to incorporate the latest on trauma. There is little question that traumatic experiences literally change our minds by altering the DNA that controls brain functions. It doesn't take a degree in science to observe the correlations between emotional vulnerability and addiction, between numerous community, peer and family risk factors including poverty, and susceptibility to addiction. The bottom line is that psychic pain creates strong motivation to not feel, to escape, to become numb, or at least to be distracted from internal despair. The long shadow of unrecognized and untreated trauma is increasingly seen as the major environmental factor affecting the rise in addictions of all types. With this new understanding of the inseparable consequences of nurture on nature and the impact of that dance from the beginning, we are finally moving away from a characterological interpretation to a long-needed biological conceptualization of the roots of emotional and behavioral illness.

Although there are many reasons why people become addicted, if the ACE Study findings are credible—and we have very good documentation that they are—adverse childhood experiences are highly correlated with addictive behavior in adolescence. From the perspective of the traumatized brain, substances and experiences that would not otherwise be attractive are viewed as a safety net rather than a trap. For untreated trauma survivors, drugs may offer the only relief in sight. For some, the initiation into drugs may begin with drugs originally prescribed by a physician for anxiety, pain or learning disabilities like ADHD. For others, the route is through street drugs made available through peers. But for the adolescent whose normal anxiety or depression is intensified by trauma, the attraction to such substances is amplified.

When genetic vulnerabilities are added to the equation—for example, when genetics exacerbates the effects of alcohol on the brain or dysregulates the balance between limbic and cortical messages in decision-making—addiction is even more likely to serve as a bridge to adult disease. Smoking is an example. Among white smokers of European descent, if a certain genetic variation is inherited from both parents, there is a 23 percent chance of getting lung cancer—a 70 to 80 percent greater chance than the odds for a smoker without the genetic vulnerability. If the inherited genetic variant comes from only one parent, the chance of developing lung cancer is reduced to 18 percent—about one-third higher than for people without the variant. For those who smoke but lack the genetic variation, the chance of getting lung cancer is only 14 percent. For people who neither smoke nor have the genetic variant, the chance of developing lung cancer is less than 1 percent. The genetic variant is not a direct cause of lung cancer and may never manifest unless the individual smokes.

According to Christopher Amos, professor of epidemiology at the MD Anderson Cancer Center in Houston, gene variations first make a person more likely to become dependent on smoking and subsequently less likely to quit. But he or she does not become addicted to cigarettes without making some bad decisions. It is this area of control and judgment, of impulse and modulation, that becomes a war zone for people with untreated early trauma. For those who have the vulnerable genetic variants, the battle is compounded.[36] The American Cancer Society reports that of the 44 million smokers in the United States, 70 percent say that they want to quit. Although 40 percent do quit each year, only 4 to 7 percent of that group quit for good. Drug experts say that nicotine is even more difficult to quit than heroin. Help may be on the way in the form of a vaccine under development and soon to go into third-phase clinical trials; FDA approval is anticipated in late 2011.

Clearly genes play a role in addiction, more for some drugs than others and more for some people than others. Genes can affect why we start to use, how much the drug activates the reward circuitry, why we continue to use even after we know better, and whether we will decide to quit. But at each of these junctions, epigenetics can drastically influence the outcome. In the absence of chronic stress or trauma, the genetics of addiction and of many forms of disease may lie dormant. The science in this arena is young, but exciting and full of potential and hope.

THE EPIGENETICS OF MENTAL HEALTH

Research on the epigenetics of emotional disease is so new that hypotheses and findings are announced weekly. Since the completion of the mapping of the human genome and the subsequent recognition that the findings did little to explain many forms of illness, virtually every major form of mental illness is being scrutinized for epigenetic underpinnings. In this section, we discuss some of the stronger and more interesting findings as they pertain to early stress and trauma and to the opportunities for early intervention and prevention of the most common forms.

Depression

In the many ongoing efforts to locate the genes involved in common forms of mental illness—including autism, bipolar disorder, schizophrenia and depression—all of these emotional pathologies are viewed as stemming from environmental experiences in interaction with a genetic proclivity. So some of us appear to be intrinsically wired for heightened sensitivity to difficult experiences. Life events like death, birth and divorce are hard for everyone, but such events may trigger far more than the normal sadness in those who are the most genetically sensitive. Some scientists believe that depression—which is characterized by inertia and a total inability to focus—is a physical illness typified by a reduction in the size of the hippocampus. There is a growing belief, in fact, that most so-called mental illnesses are nothing less than physical illnesses of the brain that in turn have led to psychological dysfunction.

One such researcher is Dr. Ronald Duman of Yale University, who suggests that antidepressants may do more in humans than simply treat symptoms. Duman has discovered that antidepressants administered to rats cause neuron regeneration in the hippocampus; rodents exposed to electro-convulsive therapy have shown a similar effect.[37] Progress in the search for suspected genetic variations in depression has been slow and is laced with contradictory findings. For example, in 2003, Dr. Terrie Moffitt and her husband and co-researcher, Dr. Avshalom Caspi, found that people with a specific variation of the 5-HTTLPR gene were more likely to suffer depression following childhood abuse and that having the more vulnerable variation had no impact unless the individual suffered abuse. This research was widely accepted and reported in mass publications, including the *New York Times Magazine*. Several other respected scientists built on these findings. Six years later, however, in 2009, a new meta-analysis by the National Institutes of Health (NIH) of fourteen studies found no clear correlation between the gene and depression.[38] The analysis found that the number of stressful life events were a strong risk factor for depression, but found no significant links between the vulnerable variation of the 5-HTTLPR gene and depression.[39]

It may well be that there is a genetic component to depression that has not yet been validated through multiple investigations. Clearly this is a science in its infancy, one that, like most infants, is changing before our eyes. Although research on human subjects is obviously difficult, Joan Kaufman of Yale was able to study the 5-HTT effect in a group of 196 children who were removed from their homes because of abuse or neglect. She found that, in spite of differences in the 5-HTT gene and the fact that children with the so-called vulnerable variation had the highest scores for depression, the presence of a trusted key adult greatly modified the outcomes. Children who had the vulnerable variation of the gene and who also rarely saw an adult they trusted to confide in, share good news and fun with, or get advice from had extreme depression— "off the charts," according to the report. But when children with the vulnerable variation saw an adult they trusted on a daily or almost daily basis, their depression scores closely resembled the scores of children who did not have the genetic variation. While researchers have varying opinions about the implications of this work—ranging from promoting designer drugs for children to genetic engineering—it is clear that relationships and indeed just one person can make a huge difference in the health of a child.

Another recent finding on depression is that certain variations in the CRHR1 gene, which helps regulate the stress response, can protect adults who were abused as children from developing depression. Adults who had been abused but lacked the gene variations had twice the symptoms of depression as those with the protective version. This study, funded by NIH, validated previous evidence that the stress hormone corticotropin-releasing hormone (CRH) plays a role in depression. We have known for some time that stress and trauma in childhood can hyper-activate this response system, increasing the risk of depression. But the question remains: why do some individuals have the protective gene while others do not?[40]

In November 2009, the journal *Nature Neuroscience* published what it said was the first study showing in epigenetic terms how stress in early life can program later behavior. Researchers from the Max Planck Institute of Psychiatry in Munich, Germany, studied the long-term effects of stress on newborn mice. The pups were adequately fed but were separated for three hours a day from their mothers for the first ten days of their lives. The researchers found that the separated pups produced high levels of stress hormones that led to epigenetic changes in the gene-coding for the stress hormone vasopressin, which is key to controlling mood and cognitive behaviors. The pups had poorer memories and were more likely to be aggressive than pups that had not been stressed. In humans, vasopressin increases arterial blood pressure and regulates the body's retention of water. It is also correlated with behavioral changes in pair-bonding, parent-child bonding and mood, particularly depression.[41]

Similar to the way in which diabetes is linked by in-common processes (such as inflammation) to a cluster of diseases, including obesity, kidney disease, hypertension, stroke, Alzheimer's and heart disease, it appears that depression is linked not only to anxiety—a connection that has been well documented—but also to a cluster of its own: fibromyalgia, osteoporosis and rheumatoid arthritis. Pain is a likely explanation for the linkage, but researchers are also looking for in-common epigenetic processes that would explain the correlations. For example, Hebrew University scientists reported in November 2009 that in data from twenty-three research groups in eight countries, comparisons of bone density among depressives versus nondepressed people indicated that depressed individuals had substantially lower bone density and greatly elevated activity of cells that break down bone. An earlier study reported in 2006 had already demonstrated the efficacy of antidepressant drugs in preventing and slowing bone loss. The association between bone loss and depression was stronger in women, particularly in young women, than in men.[42] The link between depression and rheumatoid arthritis, on the other hand, appears to be associated with higher levels of CRP, which is associated with inflammation and has been correlated for some time with depression. Epigenetic research on such linked diseases may uncover in-common markers that eventually lead to ways of reducing inflammation.

Autism is also linked to depression. A large, satellite-generated database for families with autistic children shows high rates of depression among mothers of autistic children. This finding is not surprising given the difficulties faced by many of these families in finding support and services. But the data from the Interactive Autism Network indicate that more than half of the mothers were diagnosed with depression before their autistic children were born.[43]

A final linkage to depression is found in people suffering from Alzheimer's disease. For six years, researchers from Erasmus Medical Center in the Netherlands followed 486 men and women ages sixty to ninety, all of whom were free from dementia at the start of the study. They found that those who had experienced depression before the age of sixty had four times the risk of Alzheimer's disease as those with no history of depression. Several studies have disproved the idea that depression was simply a consequence of Alzheimer's; in a study of 917 nuns, priests and monks, depression did not increase in the early stages of Alzheimer's, when subjects were confronting a new diagnosis. Taken together, the studies confirm that depression, particularly with younger onset, is a risk factor for Alzheimer's. Effective treatment of depression may prove to gentle the symptoms of Alzheimer's.[44]

Post-Traumatic Stress Disorder

Of all of the forms of mental illness related to trauma, PTSD is the most obvious. Symptoms can surface long after the event and include recurrent invasive memories of the trauma, debilitating irritability, insomnia, hypervigilance and extreme reactivity to sudden sounds or movements reminiscent of the traumatic event. The rate of PTSD is soaring as veterans return from Iraq and Afghanistan. Adults who suffer the most serious symptoms have been found to have a variant of a gene that is highly influenced by the experience of early trauma, particularly if it involved physical or sexual abuse.

In a study of nine hundred primarily African American eighteen- to eighty-one-year-olds, a particular gene variant, FKBP5, correlated most strongly with risk for PTSD. FKBP5 regulates the stress response system by producing protein that controls the degree of binding between stress hormones and their receptors on cells. The gene variation can cause alterations in this process, leading to excess stress hormone reactivity. Many of the study participants had also experienced severe trauma in childhood and additional trauma as adults. Dr. Kerry Ressler, co-author of the study at Emory University, says that there are critical periods in childhood when the developing stress system is particularly vulnerable to outside influences. Dr. Thomas Insel, director of the National Institute of Mental Health, believes the findings help explain why some individuals exposed to trauma on the battlefield have paralyzing and invasive reactions while others who live through the same events are better able to recover. Though both child abuse and adult trauma increased the likelihood of serious symptoms, more than twice the number of symptoms were observed in those who had survived child abuse and also had the gene variant responsive to stress. Neither element alone accounted for PTSD; both elements had to be present to result in high risk.[45]

Although many of us worry about PTSD in relation to the war veterans returning from Iraq and Afghanistan, going unnoticed is a large group of Americans who are at significantly higher risk of PTSD than combat veterans: children in foster care. In 2003 the Casey Family Programs in Seattle, in partnership with researchers from the Harvard Medical School, conducted a study of 479 adults in Washington and Oregon who had been in foster care sometime between 1988 and 1998 for a mean length of six years. The average age of the participants was twenty-four. The researchers found that they had PTSD rates that were significantly higher than those of combat veterans. Twenty-five percent of the foster care alumni met the diagnostic criteria for PTSD, compared to 15 percent for Vietnam veterans, 12 to 13 percent for Iraq veterans, 6 percent for Afghanistan veterans, and 4 percent for the general population. The recovery rate was significantly lower for foster care alumni compared to the general population (28.2 percent versus 47 percent). Mental health outcomes for survivors of foster care "appear to be disproportionately poor in comparison to the general population," including three times the rate of panic disorders, seven times the rate of drug dependence, seven times the rate of bulimia, and nearly two times the rate of alcohol dependence.[46] With over 425,000 children in foster care on any given day in our nation, these findings point to a harsh reality for children who chronically live in conditions that have worse outcomes than frontline combat.[47]

Schizophrenia

Research on schizophrenia, viewed as arising from the interaction of inherited genes with the environment, has focused on discerning either the genes or the suspected environmental factors, including maternal diet, stress, drug abuse or exposure to infections. But differing findings across multiple studies have made it almost impossible to confirm even one consistent environmental factor. Adding to the difficulty in studying this disease is the fact schizophrenia does not occur in animals. And while adoption studies and research on twins have lent credibility to the genetic argument, traditional genetic explanations fall short. For example, there are many cases in which one identical twin is schizophrenic while the other is not.

In the belief that epigenetic processes such as methylation can explain the complexities of schizophrenia, several researchers, like Arturas Petronis from the University of Toronto, are turning to epigenetics to explain the roots of the disease. Studying schizophrenia in identical twins, Petronis and his team have found that, while the healthy twin has numerous methyl tags attached to certain genes that block it from making the neurotransmitter dopamine, the schizophrenic twin typically has almost none and is producing much higher levels of this neurotransmitter.

The epigenetic model of schizophrenia can be conceptualized as a domino effect of negative events, beginning with a tiny epigenetic change that takes place in the earliest development of the embryo, or the gamete, as it is first forming. This event is not in itself enough to cause the disease; its manifestation will depend on the effects of a cumulating series of factors occurring before and after birth. Hormones, external environmental factors, and a series of random events will shape the degree of epigenetic dysregulation. Advanced paternal age is one of the suspected factors that may catalyze epigenetic dysregulation from the time of fertilization, owing to parental imprinting errors.[48] Epigenetic dysregulation may fluctuate over time, and thus the symptoms of schizophrenia may also fluctuate or even slowly recede, leading to partial recovery.

Mapping the epigenome is the long-range goal of research on schizophrenia and other psychiatric and nonpsychiatric diseases, including depression, arthritis, osteoporosis, autism, Alzheimer's and bipolar disorder.[49] It has already been determined that schizophrenia and bipolar disease share a common genetic link. A recent study of 9 million Swedes born between 1973 and 2004 looked closely at the 36,000 people with schizophrenia and the 40,500 who were identified with bipolar illness. The analysis revealed that when a child had schizophrenia, first-degree relatives—parents, siblings or offspring—were more likely to have one condition or the other. Even adopted children with a biological parent with one of these disorders were at increased risk for the other disorder.[50]

Dr. Eric J. Nestler, chair of the Department of Psychiatry at the University of Texas Southwestern Medical Center in Dallas, has proposed an animal model of schizophrenia that includes epigenetic changes in the hippocampus that result in its shrinkage, as happens in both depression and Alzheimer's. Nestler and other scientists are working to create compounds that will adjust the epigenetic mechanisms characteristic of all three of these diseases affecting the hippocampus.[51]

A final note on schizophrenia: several researchers think this disease may be linked to exposure of the pregnant mother to an infection. It appears that some families carry a genetic variation that causes a mother's immune response to a viral or microbial infection to trigger changes in the child's developing brain that ultimately result in the disease. The stretch of genetic code affecting immunity is still mysterious to researchers; most believe that diabetes, arthritis and Crohn's disease may all be sensitive to the same processes (another cluster). In each of these diseases, the body's immune system attacks the body's own cells, which may also be what is happening in the schizophrenic brain. Moreover, the genetic variations linked to schizophrenia appear to be linked to depression and bipolar disorder.[52]

On a more optimistic note, it also appears that through nurture some families may convey protective factors that greatly reduce the likelihood of a child developing schizophrenia, even where there is a genetic predisposition. Many of the common risk genes are seen in healthy people who never have the disorder. A forty-year longitudinal study from Finland focused on which family environmental factors might play a protective role in children. The results suggest that there are protective factors that reduce risk for a susceptible child by up to 86 percent, factors that take us right back to Michael Meaney and his high-licking-and-grooming mother rats. The Finnish study concludes that, if a genetically at-risk child does not experience a negative familial environment, he or she may never develop schizophrenia or, at worst, may develop a less serious form. The study recognizes, however, that even in healthy family environments, early environmental factors like exposure to prenatal stress, toxins or nutritional deficiencies may trigger problems in children genetically at risk.[53]

In his work with children and their parents, the late Dr. Stanley Greenspan, clinical professor of psychiatry and pediatrics at George Washington University Medical School, demonstrated the power of early intervention with what are still thought to be genetically linked learning disabilities (especially autism), not by solely treating the child but by working with the child and parents together. Famous for his concept of "floortime," Greenspan long contended that even in children who showed clear signs of autism or had a genetic load toward mental illness or other developmental disabilities, he could minimize trauma and the expression of genetic tendencies if he could work with the child's parents or caregivers together with the child in the child's milieu, beginning early in life. The key, he maintained, is altering patterns of relating with a focus on creating shared positive emotional states and increased meaning and synchrony in communication between parent and child.[54]

Alzheimer's

Alzheimer's disease is a gradual and progressive brain disorder resulting in dementia and death. It creates amyloid plaques, tangles and a massive loss of neurons in the brain, particularly the hippocampus. In 2010 the Alzheimer's Association reported that 5.3 million Americans had Alzheimer's, which has become the seventh-leading cause of death in the nation. As the baby boomers have aged, it has been the fastest-growing type of dementia, jumping 46 percent in just six years, from 2000 to 2006, and generating $172 billion in annual costs, not counting the value of supportive services by an estimated 10.9 million unpaid caregivers.[55] Although the majority of Alzheimer's cases surface in the elderly, it appears to have earlier beginnings. Harkening back to the Barker hypothesis, there is a growing belief that the roots of the illness begin in earliest development and result from environmental factors interacting with genes, beginning during preconception and continuing through embryonic, fetal or infant phases from stress-induced epigenetic dysregulation of key genes. Suspected catalysts include lead and other heavy metals, which are also implicated in Parkinson's disease and Lou Gehrig's disease.[56]

Researchers from MIT, in collaboration with the Howard Hughes Medical Institute, have found that an enriched environment can actually restore memory in mice with Alzheimer's-like brain degeneration. Even more surprising is their finding that the animal's memory loss could be reversed through the use of a drug. The research zeros in on a gene called P25 that induces neuronal death in regions of the forebrain and is implicated in several other neurodegenerative diseases; the expression of this gene can be catalyzed by a dietary supplement of a chemical called doxycyclin. Researchers first conditioned the mice to fear a compartment in their enclosure by inflicting mild shocks each time they entered. After several exposures, the mice would freeze upon entering. The mice were also trained to perform a spatial memory task that involved finding a platform immersed in "murky water." After eleven months of conditioning, doxycyclin was administered for six weeks. After receiving the drug, their memory on both tasks was seriously impaired. The mice were then returned to their cages, half of which were filled with toys, running wheels and various other stimulating opportunities that were changed daily for four weeks, and half of which had no added stimulation. Upon retesting both groups on the two tasks, the mice exposed to the enriched environment performed significantly better than the control animals: they froze when re-entering the compartment and successfully located the submerged platform. The mice that had received no additional stimulation neither froze when placed in the shock-filled compartment nor were able to locate the submerged platform in the murky water.

On autopsy, the scientists found no difference in the brain weight of the two groups of mice. In spite of extensive neuronal death from doxycyclin, the mice that had experienced enriched environments were able to retrieve the memories. The conclusion of the researchers was that it is the retrieval process, not the stored memories themselves, that is lost in neuronal degeneration. An enriched environment reestablished the synaptic networks necessary to the retrieval process without inducing the growth of new nerve cells or altering DNA.

Once again, this is epigenetics at work. Autopsy revealed that the enriched environment had caused increased histone acetylation in cells of the hippocampi of the stimulated mice, enabling expression of the genes that control processes underlying memory. The researchers then created a drug that catalyzed the same epigenetic process as an enriched environment, resulting in similarly improved memories in the rodents. Although there is not yet evidence that this drug would have the same beneficial effect in humans, there is hope that similar processes, including environmental enrichment, can slow the progress of Alzheimer's and other forms of dementia in humans, particularly if administered in the early stage of the disease.

Autism

There are many contradictory theories about the causes of autism. It appears that there are genetic components to the disease, although this has not been confirmed. Attempts to explain autism often point to environmental toxins, including, in spite of strong recent evidence to the contrary, the components used as preservatives in vaccines. It may in fact be that the autism spectrum blends several different diseases that will ultimately be distinguishable as we fine-tune our understanding of the underlying epigenetics. The inclusion of Asperger's syndrome on the autism spectrum adds further complexity, since that diagnosis now encompasses such a wide array of symptoms. Autism rates are hotly debated, with estimates ranging across the states from one child in 67 (in Minnesota, known for its child- and family-focused services) to one in 317 in Mississippi, where such resources are scarce. Some people believe that the rising rates simply reflect the addition of Asperger's syndrome to the autism spectrum, which occurred within the past two decades. Most experts point out that rates vary with the availability of professional resources in communities that can screen for and intervene in the problem. Much of the research is examining the linkages to environmental toxins, though clear correlations have been elusive. The CDC estimates that the current rate of autism in the United States is one in 110 and that rates have escalated since 1992 when they were first recorded.[57]

One new and yet unproven theory, from scientists at the Albert Einstein College of Medicine at Yeshiva University, is based on a study of the claim by many parents of autistic children that fever improves autism. These researchers believe that the brains of autistics are structurally normal but become dysregulated by such conditions as fever and might be reversible. They suspect that the locus coeruleus (LC) may be the source of the problem, since the LC is the one part of the brain involved in both producing fever and controlling behavior. It also secretes noradrenaline, which is key in fight-or-flight and attentional focusing. The researchers believe that stress plays a role in the dysregulation of the LC, particularly in the later stages of prenatal development. As evidence, they cite a 2008 study that found a higher rate of autism in children whose mothers had been exposed to hurricanes and tropical storms during pregnancy.[58] If their theory about the LC is correct, it means that thousands of genes may be at play and that epigenetic therapies might hold the key to healing autism. Theory co-author Mark Mahler says: "The only way you can reverse this process is with epigenetic therapies, which, we are beginning to learn, have the ability to coordinate very large integrated gene networks."[59]

Worldwide, epigenetic research is exploding. In 2007, NIH announced that epigenetics would be one of its top five research priorities for the next five years. In 2009, NIH funded twenty-two grants designed to understand "how epigenomic changes—chemical modifications that result from stress, diet, aging or environmental exposures—define and contribute to specific human diseases and biological processes."[60] In 2008, NIH put $190 million into the Human Epigenome Project. In September 2009, NIH added another $62 million to fund more than twenty additional studies under the Epigene Roadmap Program. Understanding cancer is one long-term goal, but scientists are also looking at an array of other diseases, including autism, hardening of the arteries, glaucoma, aging, asthma and abnormal growth and development. Dr. Francis Collins, director of NIH, is optimistic, saying, "Epigenomics represents the next phase in our understanding of genetic regulation of health and disease."[61]

In Europe, public and private laboratories are working together to map the range of intracellular molecules that influence the human genome. Pharmaceutical companies are searching to develop new epigenetic drugs to treat an expanding array of diseases, including schizophrenia, Alzheimer's, sickle cell anemia, lymphoma, and breast and brain cancers. The challenges in epigenetic research are daunting, and not the least of those challenges is the fact that the epigenome is exponentially more complex than the genome.[62] Nonetheless, this is the new frontier, and the possibilities stretch beyond imagination.

OUTTAKES

My husband and I live high in the hills of West Portland. After both he and I received cancer diagnoses only weeks apart (we are now both cancer free), we both questioned the potentially carcinogenic role of the huge radio towers near our home. While researching this chapter, I was telling him one evening about some of the findings concerning the generational transfer of epigenetically altered genes that might be stealthily handed down for several generations. "Makes it pretty hard to blame it on the towers," he said. We laughed. But in fact, it may have been the towers, along with diet (diet soda for me, hot dogs for him), early trauma (in my case, not his) and toxins that we have little idea of—a cumulative soup adding up to carcinogenic levels for each of us. It is so tempting to think in terms of a silver bullet, as the media loves to do, pinning the blame on a food, a cell phone, a pesticide or whatever. The reality is that each of these and more may lie at the root of many cancers and other chronic diseases. Together, all that we eat, all the we drink and feel creates the chemical stew around our genes that together shape out health.

The question is what to do. Where can we begin to have the most impact? The evidence increasingly points to the beginning when life is just forming and is most vulnerable. We may all be born equal, but we aren't born the same. We each come into the world already reflecting our own unique level of sensitivity, already expressing our own experiences in the womb along with those of our progenitors. Our childhoods, even for identical twins with the same parents, are not identical. The toxins that may cumulatively shape our health are not just "out there." They are also within our experiences—and those of the people who have come before us and those whose lives we entered before we had words. When trauma or lack of nurture has been part of that equation, though it may be without conscious memory, it is not without consequence. To quote Dr. Scaer: "All the cells of the body are administered by the brain." The footprints of experience—positive and negative—register in tiny, microbiological alterations on our raw selves at the cellular level. And those footprints tend to intensify as we age, their indelible impression reflecting in our health. The trick for us as adults is to discern their imprint and to adjust our paths accordingly. But for the babies in our lives, the best antidote for trauma and the surest path to health is the consistent attuned interaction with someone who loves the child.

CHAPTER 8

Security Blanket

The Biology of Secure Attachment

WE TEND TO THINK about challenges to infant development categorically. We worry about accidents and illness, a surgery the child must undergo, the effects of divorce or separation, inadequate child care. Each of these challenges—and others—can pose a threat to the emotional health of a baby or toddler. But it is fundamentally the security of the attachment relationship that mitigates or exacerbates the impact of these experiences. We have all had "little traumas" in our childhoods. It may be that you simply had your tonsils out or had to go to a school you hated. You may have been adopted, lived through a divorce, or lost a sibling, a beloved grandmother, or a favorite pet. The issue is not just "What happened?" but also "What meaning did it hold?" How often did bad things happen? At what age did they occur? And what kind of nervous system did you have to begin with?

But most profoundly the impact of trauma is ruled by the question of "Who?" Who was there with you and how deeply did you feel connected, loved, valued and supported by that person? Just one key relationship—just one person who is available to the child over time, who sees the baby as valuable, and who communicates that feeling—can make all the difference in how later stress or trauma affects that child's future.

Imagine for a moment being in love—more in love than you can remember. Imagine that her face, her attention, is your first thought when you wake up and the image you seek to help you relax into sleep at night. Others can come and go, but it is she who looks and feels and smells and tastes familiar; when she is with you, the world feels safe and interesting. She is the other part of yourself. Her rhythms, her voice and her touch shape your days. When she is not there, you watch for her and feel much better when she returns.

Now imagine that you can't talk or even form meaningful syllables, that you can't walk or pick yourself up, that you can only open your eyes and make a few sounds and movements to let her know that you need her. Your body knows that your very life depends on her—her caress, her recognition of your wordless cues, her knowing when you are hungry or hurt or uncomfortable. She is your world. If you know she will be there, you relax and can enjoy the world around you. Together you create a dance in which you know each other's rhythms so that it feels like you are one. "Attachment" at this stage in life is essentially the professional word for what most of us call "love."

LOVE LOST

But what happens when she is gone? When she suddenly doesn't come back for the dance you have known since birth? You see faces, but they aren't the right face. You hear voices, but you keep listening for the one you know. You look for the familiar features of her face, you wait for her smell, her touch. You are picked up over a strange shoulder and then laid down; you search again for the face that has always been there when you are in this position, but you can't find her. Though that wet thing around your middle comes off and a dry one feels better, the touch and the smells and the textures are all wrong. The nipple, the rhythms and positioning of your body—none of it is right. You can't settle. You can't get comfortable. You pull away and turn your head to see if she's coming. You see some familiar things around you, and you look for a while at the colors and the movement, but then you automatically picture her face behind your eyes and you stretch to look again for her. Finally you sleep. But as soon as you wake up your first thought is to find her again with everything you have. Her picture is in your mind. You writhe and you arch your back to move away from the other who is not her. After a while, you feel frustrated and afraid and your stomache hurts and you begin to cry a little and then a lot. Your eyes and your throat ache. Your heart beats fast, and it's hard to catch your breath, and you want to go back to sleep. Nothing feels right. You look and look and she isn't there, and you don't want to see the rest.

This is the baby when attachment is lost for an extended time and is more than a baby is prepared to handle. To most adults the traumatized infant may look "fine," a bit fussier than normal or slightly subdued. The baby may either eat almost nothing or sleep and eat more than usual. Some children are highly reactive and hypersensitive: they cry, startle at every sound, kick and are hard to comfort. Others are frozen and are often seen as exceptionally "good" babies. Initial restlessness and fussiness may give way to fatigue, the baby may give up, and very real depression may set in.

Unless we are looking closely and are fairly educated about baby cues, it is easy to overlook infant trauma. A baby may seem to adapt and roll with the punches. After all, babies are malleable. Unable to verbalize the strong feelings taking place internally, their language is behavioral and physical. Because we have known so little about emotional trauma in infancy and are unused to paying attention to these subtleties, much of a baby's grieving or depressive behavior may go unrecognized and be dismissed in a too busy world. For the baby there may be no greater trauma than failure to achieve and maintain a secure sense of connection to a caring adult who is able to help regulate the baby's fundamental physiology. But recognition of these dynamics remains a huge issue even for professionals: diagnosis of profound emotional distress in infants and toddlers has only been formally available to clinicians for the last decade, primarily through Zero to Three, the National Center for Infants, Toddlers and Families. This information is not addressed in the current edition of the official *Diagnostic and Statistical Manual of Mental Disorders* (*DSM-IV*), which is the primary diagnostic reference for mental health practitioners.

This doesn't mean that no one but a mother can care for a child or be part of an attachment relationship. Dads and grandmothers and nannies and other alternative caregivers are optimally a strong part of the attachment circle and sometimes the primary attachment figure for a baby. And even though the majority of the research has been on mothers, we know that fathers are increasingly assuming this role of caregiver in today's families.

As the child's world expands, relationships are best built through gradual daily exposure in the company of the primary attachment figure so that comfort is associated with the primary figure and care is transferred in a familiar and gradual manner. While it may be traumatic at first for a baby to transition to an alternative caregiver, such a little trauma is not likely to have a lasting negative impact if the baby has had enough time in infancy to build a solid relationship with a consistent attachment figure and if the primary figure returns in a reasonable time and resumes a connected relationship. There is evidence that short periods of separation make for resilience in young children.

Major breaks in attachment and the resulting trauma to the baby's nervous system are sometimes unavoidable. Tragedies occur, and even though life need not be forever bleak for those of us who lose the original attachment that is so protective, there are repercussions that are not likely to go away by themselves. There are no universal equations; each of us is different. The questions that seem to matter most are: For how long was the relationship broken? Who else was there? What happened before? Was there any history of prior trauma? How warm, stable and capable of attunement was the attachment figure or the adoptive caregiver?

"IN LOVE"

Let's return to the analogy of being in love. This time, think back for a moment to a time *before* you were really in love in a mature sense. Try to remember your first huge crush or blind adoration of someone you really wanted to be close to, someone you longed to have pay attention to you. It may have been a person of the opposite sex or the same sex, someone your own age or someone older. You may have been a child or a teenager, but when this person appeared you had your heart on your sleeve—and no defenses. You may have felt rubbery: his or her gaze was so powerful that you had to look away just to get a hold of yourself. Perhaps you were energized and felt impulsively compelled to show off—your arms and legs seemed to have a life of their own. You thought this person could read your mind. You scanned his or her face for signs of acceptance or rejection, and you were so keenly tuned in to this person—so wired, so vulnerable to their slightest gesture, that a glance, a smile, a scowl, a turning away, or the smallest comment or change in tone could make your day or destroy it, in a heartbeat.

Most of us have had this experience, which is essentially a short-lived regression to a time long ago. Although such experiences are (fortunately!) usually brief, for a time we are reduced to our baby selves, to that core of early vulnerability to the other. The memory of the original experience wasn't recorded in words. But the state is somehow familiar. Research suggests that how it went before you can remember—usually with your mother—has great bearing on how you feel and behave when you encounter these feelings in an adult relationship.

For the infant, as for adolescents and adults, it takes time and many interactions before infatuation develops into a fully attached "in love" relationship. But from the beginning we are primed by intense needs for "the other." Once attachment is securely accomplished, children, like baby monkeys, are more likely to explore their environment, reflect curiosity, be persistent in complex tasks, be less fearful of change, and show less frustration while solving problems. They are more comfortable and cooperative with peers and less likely to respond aggressively. The quality of the attachment relationship with the primary caregiver has far more bearing on the child's cognitive, emotional and physical health than the alternative child care experience, which tends to get far more press. Countless studies have demonstrated that secure attachment is the best defense against later social and behavioral problems, including both aggressive behavior and victim-prone behavior.[1]

Contrary to what the term "bonding" implies, the attachment relationship is not instantaneous. It does not just happen at birth. We may feel an immediate love for an infant at birth, but attachment is like love later in our lives: the real thing takes time. Secure relationships at any age take time as well as opportunities to be together, to attune to each other, to learn each other's meanings, to engage in reciprocity and shared emotional states, but also to endure breaks in connection and to mediate the resulting anger and disappointment, and finally, to repair and reconnect.

"Attachment"—or the lack of it—is all around us, reflected in our daily encounters not just with our parents or our children but also with friends and neighbors and even to some degree with the people who repair our cars or perform our surgeries. Attachment affects our days and nights, our social structures, our security and most profoundly our bodies. We get sick and we get well inside of relationships. And how those relationships will tend to go for each of us has everything to do with our first ones—those we remember primarily not with words and thoughts but with feelings and sensations stored within our bodies.

In pragmatic terms, an attachment is a tie or fastener that attaches one thing to another. And so it is with people. Significant emotional bonds (along with legal ones) connect humans within families, cultures and nations. This felt process of gradually learned emotional connectedness to other people—or the failure of that process—generates stunning physical consequences that affect not only individual well-being but also familial, community, national and world health. We embody our connectedness, or the lack of it, even though the creators of the template we draw from may be long gone, along with any conscious memory of the process—because this all begins with babies.

SELF-REGULATION

Most of us have a fundamental procedural memory of how we feel when we are feeling "all right," "good," "normal" or most like ourselves, physically and emotionally. We also recognize when we are not there. Our brains regularly scan our bodies and minds to note how we are feeling in order to make the necessary adjustments that re-establish equilibrium. We reflect consciously on how much sleep we have missed or how much food we have consumed. Whether we're talking about physical problems, like a stomachache, or emotional discomfort, like being anxious or depressed, we often describe these conditions as feeling "off-kilter," "out of it," "off track" or "below par."

The skill—or perhaps the art—of balancing how we eat, sleep, work, play and connect with other people is foundational for all kinds of health: physical, relational and emotional. None of us were born with this capacity. If we have it, we had to learn it, at least at the most fundamental level, from someone else. For most of us, this learning began with a parent, typically a mother. For all of us, our expectation of either comfort or chaos from the world outside and the consequent shaping of our basic emotional rhythms were substantially built inside of early relationships before we entered school, and much of it before we could talk. The professional term for this set of skills is "self-regulation."

These are the tasks of the infant, and they are essential for survival: to attach to an adult protector and gradually, with adult assistance, learn to regulate the basic physiological processes that secure survival. Any experience that gets in the way of these fundamental needs is perceived by the baby's body as life-threatening and generates the fight-flight-freeze response. The reverse is also true: secure attachment provides trauma protection. But how? What is the biology behind this observation?

THE SOCIAL ENGAGEMENT SYSTEM

To understand the role of attachment in health, it helps to know another major piece of the autonomic nervous system. In earlier chapters, we looked at how the vagus nerve throws the switch to activate the freeze response in trauma. But it plays another role as well—one that can save the day. Dr. Steven Porges of the Brain-Body Center at the University of Illinois has uncovered an expanded role played by the vagus nerve in connecting our experiences, specifically our feelings, to our health.

The vagus nerve is basically a tube extending from the brain down through the neck and trunk into the lower groin, serving as a conduit for various nerve fibers to connect the brain to several vital functions. It consists of two parts: the older, more primitive *dorsal* vagus—the one involved in "voodoo death"—and the more recently evolved *ventral* vagus, sometimes called the "smart vagus" because it is insulated (myelinated) for more efficient transmission of nerve impulses. The ventral vagus lies above the diaphragm, while the dorsal lies below it. The dorsal vagus extends from the brain to all of the viscera in the lower body, enabling our heart, lungs and digestion to slow way down in extreme danger. The ventral vagus, however, is linked to the cranial nerves that control facial movements and vocalizations and is involved in facial gestures and listening.

Without conscious effort, our nervous systems are constantly monitoring risk. The detection of risk or safety immediately triggers corresponding internal changes in respiration, heart rate and visceral organs.
Porges's contribution has illuminated the role of the ventral vagus nerve
in providing this two-way communication between the brain stem and
the visceral organs. Higher brain structures signal the brain stem, and
the vagus mediates the responses, creating major physiological shifts in
bodily states that in turn influence our emotions. While emotions shift
bodily states, bodily states also shift emotions.

In response to threat, we have an automatic, hierarchical and highly
linked set of responses. First, our newest system—gestures, tones of voice
and facial expressions—allows us to self-soothe and create safety in the
company of people we trust. This is a sure way to relax internal states, eating and smiling, laughing, employing muscles of the head and face that
signal through the ventral vagus that all is well. Our eyes widen in attention, the muscles of the middle ear tighten to attend audially, our
heads and necks turn comfortably, our vocal chords and larynx relax. Relating to other people is a powerful way to calm the body for both babies
and adults. As long as this is working, heart rate is normal, and blood
pressure is optimal; not only are we open to connecting, but our organs
are available to physical growth and restoration.

If fear is detected, these same circuits reverse the message, engaging the
HPA axis and shifting our defenses from the parasympathetic ventral
vagus nerves of the head and throat into full mobilization of the sympathetic system. Our world shrinks to focus on the threat, and our capacity for fluid social behavior is lost. At this point we have limited sensory
awareness and reduced ability for analysis or planning. The chemistry of
mobilization is coursing through our veins, and all of our organs are receiving the message: red alert! Normal metabolism stops. Then, if fight-
or-flight is not enough to reduce the threat, if we are still helpless and
powerless in the face of the threat, the dorsal vagus system, our end-of-
the-line parasympathetic response, takes over: we freeze, becoming dis-
sociated and immobile. The freeze state is also calming, but if summoned
for too long it brings on the natural anesthesia of incipient death.[2]

This is how we react to threat: we start with our newest system first
(the ventral vagus), and when that doesn't work we revert to our older sys-

tem (the HPA axis), and then, if equilibrium has still not been restored, we resort to our most primitive system (the dorsal vagus). This all takes place in a matter of seconds, and one system may override another. For example, if a baby is howling for food, even though his or her HPA axis may be tuning up, mother's voice, touch, repositioning and suckling of the baby will immediately engage the ventral vagus system. Activity of the muscles of the mouth as the baby nurses, eye contact, the sound of the mother's voice and vestibular movements trigger the secretion of oxytocin, a chemical message to the hypothalamus to dampen the HPA response. As the infant suckles, she will calm and facial expressions will soften and relax, limbs will quiet, little eyes will close as the baby experiences comfort and returns to homeostasis. By five or six months of age, the baby will work hard to engage her mother's face, pulling away from the breast or bottle to get eye contact for further reassurance or simply increased validation that all is well. This "social engagement system" is all about contingent exchanges between faces, voices and gestures, the language of one "right brain" to another. If it works, it dampens the activation of the sympathetic nervous system and HPA. There is no need for fight-or-flight (HPA) or for freeze. But for the social engagement system to adequately accomplish the mission in babies or in adults, the individual has to feel safe; otherwise, the system goes offline and HPA takes over.[3]

What does "safety" mean for a baby? Detection of safety—for example, being aware of the familiar versus the unfamiliar—is not under conscious control and cannot be turned off or on. This is essentially radar for survival. The newborn is vulnerable to the shifting tides of raw and erratic circuits in key systems, including gastrointestinal and sleep/wake cycles that are still unregulated. The perception of safety for the baby is dependent on an attuned adult system to maintain homeostasis for the first years by consistently meeting basic needs until circuits are built that allow the child to regain homeostasis by herself.

Predictable responses by an adult to the baby's attempts to engage in this vital dance are experienced by the baby as calming and pleasurable, dampening the HPA and gradually creating a sense of trusting connection. The social engagement system builds the best and first line of defense against trauma. To the degree that this is not available—if, for example, the adult is unfamiliar or is unpredictable or inflicts pain—baby's radar goes into high gear. Social engagement will be minimal and HPA will prevail. Trust in relationship may be sacrificed. At the mercy of their chronically stimulated stress response systems, these are the children most vulnerable to trauma.

In the 1950s, researchers began to recognize that secure attachment is far more than just a nicety. Harlow's monkeys and children left in orphanages after the war demonstrated that individual adult attention—especially touch—is so essential to physical, emotional and cognitive growth that its absence threatens survival. In the 1960s, John Bowlby's "attachment theory" pointed out that constructive emotional nurturing is essential to adequate physical as well as social and cognitive development.

This was still a hard lesson for those hoping to remediate the life course of many Romanian orphans after the fall of communism in Romania. Dr. Charles Nelson, a Harvard neuroscientist who studies early brain development, filmed three boys who grew up in these emotionally deprived environments. In the film, the boys, whose legs do not touch the floor in school-size chairs, look to be five or six years old. In fact, they are each between fifteen and twenty years old, testimony to the physically stunting effects of profound emotional deprivation. Nelson explains that while the boys got enough to eat, without adult attention—love, by any other name—they didn't produce enough growth hormone because their bodies were conserving energy for brain development. For these boys, there was so little attention and stimulation to nurture their brain growth that their IQs were forty points below average for their age. Genetics don't account for this outcome. In a September 2006 interview on National Public Radio, Nelson said: "If you're exposed to the wrong experiences or you have no experiences, such as what occurs in deprivation, then the brain kind of gets miswired. The concern we have from a neuroscience perspective is if it is miswired early on for an extended period of time, it may be very difficult later on to rewire it."[4]

Nelson and his colleagues began the Bucharest Early Intervention Project to examine the effects of early experience on brain and behavioral development. One hundred and thirty-six institutionalized children ranging in age from three months to thirty months were divided into two groups. In one group, sixty-eight children have been randomly assigned to remain in one of the six orphanages. In the other group, sixty-eight children have been randomly assigned to foster homes in Romania. The plan is to compare the development of the children placed in foster homes to that of the children in the orphanages as well as to that of a third group of children who are being raised at home by their parents and who experienced no early breaks in their care. The children will be followed and assessed on several measures through their seventh to ninth years.

The researchers' immediate goal is to prove to the Romanian government that foster care is a better option for all young children than institutionalized care. They hope to train and provide specialized support to a large cadre of foster parents so that all children are raised inside of families. But the project is also an opportunity to illuminate the effects of early experience on emotional, social and cognitive outcomes as well as physical growth. The Bucharest children are now five to seven years old, and researchers have already confirmed that age matters. The younger the child when he or she was placed in foster care, the better the gains in IQ and overall development. Touch, talk, encouragement and cheering a child's progress make a huge difference.

Five-year-old Florin was placed with foster parents when he was eleven months old. At the time of his placement, Florin couldn't sit up, a milestone accomplished by most babies at around six months. After just one month of foster care, Florin was sitting. By fourteen months, he was walking! Nelson says that the IQs of babies placed before twelve months have risen more than ten points. They also are showing lower rates of depression and anxiety. Some things, however, have not changed. Rates of ADD have not improved, and though the children grew in height and weight, their heads stayed small for their ages. Their rates of brain activity are not as vigorous as those of the children who had not been institutionalized.

Researchers believe that there are sensitive periods in early life for specific aspects of development. The first year of life is a critical period for head circumference—growth in the actual size and function of the brain. The second year appears to be critical for language. Whether created by birth, adoption or foster care, we already know that children do best in families. But Nelson's project reveals that not just any foster or adoptive family will do for children who have been deprived early. Normal kids who have never been traumatized by separation and deprivation benefit from a typical home—day-to-day family life that provides the basics of touch, talk and stimulation—but babies and young children who have experienced chronic deprivation require, in Nelson's words, "super-duper" foster families to catalyze their progress.

This all takes time. Even for normal babies under the best of circumstances, building social engagement skills takes time. The process of learning how to "talk" before words takes time when time is money for families struggling to provide for their children. It takes time in a culture that values the tangible benefits of an extra paycheck. It takes time when child care is underfunded and it is a struggle to provide wages that support competent caregivers. It takes time when a society is convinced that the most important early learning happens with "Baby Einstein" on a computer screen rather than through engagement with the mother's face, the most important learning of all. When early development and social engagement capacities are challenged, Porges reminds us, whether evidenced by seemingly minor symptoms like chronic gaze aversion or by more serious ones such as reactive attachment disorder or autism, a child can't just "turn it off. We have to understand that these feelings are physiological events triggered by specific neural circuits, and we need to figure out how to recruit the neural circuits that promote social behavior."[5]

It is easy to both over- and underestimate the brain's potential. We know that the human brain remains plastic and is able to compensate for many types of damage. But there are also critical periods when certain functions may be forever lost. Dr. David Hubel and Dr. Torsten Wiesel merited a Nobel Prize in the middle of the past century for their research showing that, by four months of age, babies totally deprived of vision will be blind regardless of ensuing surgeries. There are some things we cannot retrieve.

In *Ghosts from the Nursery*, we told the story of Ryan, a baby who came to his family through adoption. He had been born via cesarean section to a young unmarried mother. Because Ryan had an irregular heart rate at birth as a side effect of the anesthesia administered at his delivery, the adoption agency decided at birth to "hold" him for his six-week checkup before assuming all was well. Ryan was placed in foster care until his thirteenth week in a home that was caring for nine other babies under age three. When the adoptive family first saw Ryan, they were immediately impressed: he was a big beautiful boy with red hair, just starting to curl. But his head was covered with cradle cap. He had oozing infections in both ears. Nose to toes, his body was covered with infant acne, his bottom aflame with a bleeding diaper rash. But most troubling, Ryan would not let himself be held without arching his back and trying, it seemed, to throw himself out of their arms. He would not look at a human face. If the parents tried to gently use a finger to turn his head toward them, Ryan would turn back immediately, preferring to focus on a bright light or a shining object. When he cried, he did not find pleasure in being held. Only a bottle and a flat space to be alone seemed to comfort him. It was years before his parents learned the truth about Ryan's early months. He had lain on a crib mattress with little human touch or interaction. He was fed regularly by bottles propped in his crib, so food was his main source of pleasure, an association he has never outgrown.

If you were to meet Ryan now, his differences are not immediately perceptible. He is still a handsome boy, friendly and well mannered, intelligent, with good initial social skills. But you soon sense that something is different about him. He has been diagnosed with Asperger's syndrome along with anxiety and depression and has never fully outgrown his isolation and unsureness about other people. He is withdrawn and is easily overstimulated around people. He becomes anxious and lacks confidence in a social group, and has a high need for solitude. Like the baby Ted Kaczynski, this was a child who experienced an early interval of intense relational trauma and probably prenatal maternal stress as well. From all that is known, in spite of the fact that both Ryan and Ted Kaczynski lived most of their lives with available, competent and loving caregivers and each went to college and is intellectually bright, both were scarred by a period of chronic emotional deprivation at a pivotal time in their development.

Ryan and Ted Kaczynski have little in common except their respective deficits in their social engagement systems. The inability to accurately detect safety in the environment and to judge the safeness of other people is central to many psychiatric disorders in children, including anxiety, depression and reactive attachment disorder. These children often show signs of dysregulation of the ventral vagus, such as difficulties regulating visceral states and the heart. Their faces may appear blank, their features are unanimated, and they often have limited motor control of the muscles of the head and face. Anxiety is common. Children diagnosed with reactive attachment disorder, whether they are inhibited and unresponsive or indiscriminately affectionate, also have an impaired ability to evaluate risk in the environment. And as epigenetics peels away our reliance on DNA to fully explain several psychiatric diagnoses, science is recognizing the role of the biological impact of early experience in many of these conditions.

Dr. Porges is particularly interested in the autism spectrum, which manifests as a severe dysregulation of the social engagement system. He has developed the Listening Project, which uses computer-altered acoustic stimulation to "challenge and exercise" the muscles of the middle ear to enable autistic children to isolate and tune in to the human voice. Because the muscles of the inner ear are linked to the muscles of the eyelid and heart, the intervention not only reduces listening problems but increases eye gaze, improves auditory processing, and calms children by increasing the influence of the ventral vagus on the heart. An initial group of about one hundred autistic children listened to computer-altered sounds filtered to mimic the frequency band of the human voice. Eighty percent of the children in this double-blind randomized study showed immediate improvements lasting more than three months. Porges has enabled at least two hundred Listening Project participants to better discern the human voice from background sounds.[6]

ATTUNEMENT, MIS-ATTUNEMENT AND RE-ATTUNEMENT

From birth on, infants use their expanding abilities to engage with their social environment.* Touch, smells and sounds are critical. Although they

---

*Maternal references predominate in this section because the research typically involves mother-baby dyads. However, "father" is equally pertinent in most examples when he is the primary caregiver. Dr. Allan Schore, a world-renowned expert on attachment theory, notes that left to their preferences, babies in their first year of life naturally prefer the female adult—her voice, smell and rhythms. But during the second year this preference expands to include a nurturing father or caregiver in the child's vote for "most adored person in my world."

can see from birth, at the end of the second month there is a critical blossoming in the visual cortex. Mother's face becomes the most electric source of stimulation in the baby's world. The infant is intensely attracted to mother's eyes and will attempt to follow and regain the dynamic charge of intense mutual gaze that in turn attracts the mother. Visual magnetism for each other is evident in normally developing relationships. For both mother and baby, face-to-face exchanges generate the chemistry of pleasure, enhancing dopamine. Social play emerges as they begin to regulate each other's arousal in shared pleasure. At first the mother leads the dance by synchronizing her affect with her infant's cues; over time each adjusts their responses to the other. These are opportunities to practice the coordination of biological rhythms. Physiological monitoring of the bodies of a secure mother-baby pair reveals that over time their sympathetic and parasympathetic responses begin to mirror each other's: heart rates, respirations rhythms and periods of activity and inactivity entrain, testimony to attunement. As most sleep-deprived new parents will attest, this synchrony takes time and energy in the beginning, but gradually becomes more fluid as the baby's capacity to communicate becomes clearer. Both partners are learning, co-creating an interaction tailored to just the two of them. The mother is the biobehavioral regulator of the baby's states. The more attuned the mother, the more she comfortably allows the baby the opportunity for disengagement—looking away to modify his excitement—and the more she reads and matches his cues for re-engagement.

But there will be misses and temporary breaks—a misreading of cues by mother or a tired baby who is unresponsive to her best efforts. This is all part of the dance. Now the two have an opportunity for re-attunement— a coming back together that strengthens the connection as well as their confidence that they can do it again. The child is learning that repair is possible and that a break in "oneness" does not mean all is lost. Here is our first learning about relationships—that they are uneven and have to allow for two brains that cannot always be in perfect synchrony. But we can repair and reconnect. High rates of attunement and repair are a good predictor of secure attachment.

Dr. Daniel Siegel, author of *The Developing Mind* and a passionate speaker on attachment, explains that this regulatory dance between the baby and the primary caregiver is the foundation for both how the baby subsequently accomplishes self-regulation of his physical and emotional states and how he later relates to other people. Siegel believes that, on a brain-based level, the attuned parent is facilitating the spontaneous flow of information between the left and right hemispheres of the child's brain, allowing integration of linear, logical and verbal communication with nonverbal gestures, such as tone of voice and facial expressions. Essentially, the baby is receiving a fundamental course in "reading" not only his own internal states and another person's response but also the subtle facial and behavioral cues of the mother—building a filter for future intimate relationships. This dance of "contingent collaborative communication," Siegel says, is at the core of successful attachment between any two people throughout life:

> This process perhaps is best seen as a form of "resonance," defined as the mutually influencing interactions between two or more relatively independent and differentiated entities. . . . Resonance allows two systems to amplify and co-regulate each other's activity. . . . The integrating experience of resonance also gives rise to a sense of spontaneity and creativity when it occurs between two people. Such vibrant connections between minds can be seen within various kinds of emotional relationships, such as those of romantic partners, friends, colleagues, teachers and students, therapists and patients, and parents with children. Two people become companions on a mutually created journey through time.[7]

During mis-attunement, the relationship between the parent and child looks more like a cha-cha than a waltz. Baby's stress is evident in an unmistakable display of negative emotion, crying, and inconsolability, as the baby experiences a departure from or a failure to achieve internal balance. Although temperament differences and certain situations (like teething or illness) may also have an influence, it is not uncommon for a seemingly innocuous event—like setting baby down to answer the telephone—to generate huge wails when the baby is suddenly overwhelmed by the loss of parental proximity. She may cry and generate a series of internal reactions that place her in fight-or-flight, with all internal resources temporarily committed to essential survival functions. While an occasional incident of this nature is not problematic and some are essential, chronic mis-attunement in infancy can result in the sacrifice of potential growth.

WAYS WE ATTACH (BEFORE EIGHTEEN MONTHS)

In the 1970s, expanding on the work of John Bowlby, psychologist Mary Ainsworth researched the effects of attachment on behavior. Her groundbreaking "strange situation" study examined how infants between twelve and eighteen months respond to being left alone, comforted by a stranger, and then reunited with their mother. Ainsworth identified three different types of attachment patterns—"secure," "insecure/ambivalent" and "insecure/avoidant." In the 1980s, Ainsworth's student Dr. Mary Main added a fourth category, "insecure/disorganized."

We all form attachments; the question is, what kind? Attachment categories reflect the degree to which very young children expect to emotionally interact with and be comforted by their caregivers. Separation from the primary figure, including emotional unavailability, causes alarm for infants. How the child then draws upon the relationship with the caregiver at times of distress, their ease and interest in exploring the environment when that person is around, and the degree of pleasure the child and their caregiver take in each other's presence—all speak volumes.

Classifications

Attachment classifications range along a spectrum from "secure" to the three types of "insecure" mentioned earlier, the most extreme of which is disorganized attachment. Securely attached children are "hooked in" to their caregiver in play, often affectionate, and comfortable exploring their environment with the caregiver nearby. These are the children who have learned to trust that their caregiver will be there when needed, seek the caregiver when distressed, and are easily comforted. The key to recognizing secure behavior is the way the child goes to the parent for calming: he or she can take comfort from another adult in the parent's absence, but clearly prefers the caregiver. Securely attached parents tend to be open about their own histories and can share a coherent narrative about their own attachment. For decades we have recognized the role of a securely attached relationship in preventing cognitive and emotional challenges or intervening effectively when they do arise. What we haven't known until recently is the protective impact of attachment on physical health. Shared positive emotional states—play, joy, contentment—affect physical well-being.

Children in the "insecure/ambivalent" or "anxious" category tend to both seek contact and to resist it when it is offered. They have problems freely exploring their environment, lack confidence, and are excessively troubled by separation from their parents, resorting to regressive behaviors and meltdowns when they want affection. They are often hypervigilant, unsure of what to expect. Parents of anxious/ambivalent children are typically inconsistent and often mis-attuned, alternating between appropriate and neglectful responses when the child is distressed.

Because their signals have been either ignored or rejected, children in the "insecure/avoidant" category believe that their caregivers will be unavailable and unresponsive to their needs. Parents of avoidant children typically discourage crying and put a strong emphasis on independence, relying on distraction, redirection or irritation when a hug would be a more constructive response. In the absence of reassurance, the child learns not to look to the parent, often ignoring or turning away from the parent's attempts to engage them. Avoidant infants learn to inhibit strong emotions, especially negative ones, and avoid emotionally charged interactions, resulting in little shared emotion with the parent during play and almost no visible response to the parent's departure or return. Mothers of ambivalent or avoidant children tend to be dismissive of their own attachment experiences, glossing over their own childhood in idealized and nonspecific terms; they are uncomfortable making connections to their own history and the relationship between that history and their current behavior with their own children.

"Disorganized attachment" is the most troubling of the four categories. These children may display signs of trauma when the adult returns, such as rocking or freezing, and they have no organized strategy for adult engagement. Their signals toward the parent are often contradictory and disoriented, such as approaching the parent with their back turned. Disorganized-attached children demonstrate behaviors associated with fight-or-flight, and both the child and the parent can be dissociative— seemingly in a fog or detached from reality. The behaviors of parents of disorganized children are similar to those of their children: they are alternately fearful and frightening, intrusive and withdrawn, often negative, and prone to both abuse and neglect. Approximately 80 percent of maltreated infants may be classifiable as disorganized, compared to 12 percent in nonmaltreated samples. The link between disorganized attachment and child psychopathology is well established. Many believe that this form of attachment presages borderline personality disorder.[8] Researchers and clinicians increasingly recognize disorganized attachment as early emotional trauma, the outcomes of which surface as the diseases reported in the ACE Study.

In evaluating attachment classifications, researchers focus on how the child draws upon the caregiver for emotional support in both a typical and a challenging situation—for example, observing the pair playing together and noting what happens when the parent leaves the room, leaving the child with a stranger. There is a subtle emphasis on the contribution of the caregiver in the relationship, and often in the research the child's behavior is viewed as primarily a reflection of what the parent does or doesn't do in response to the child's signals.

Neurobiologists working with brain-based differences and developmental pediatricians who see these children along with the temperament theorists recognize that babies come into the world with genetically imbued differences in how they respond from the time of birth. This school of thought gives equal emphasis to the child's contribution to the attachment relationship. Having given birth to three children and adopted one, I know that this information is essential to adequately assess the dance between parents and children. Babies begin their journeys with many brain-based differences, the majority of which can be subtle in earliest development. Although how parents respond to those differences has a major influence on the child's behavior, it is far from the sole factor at play, and recognition of this reality is central to intervention. The adult is clearly the lead in shaping the relationship. But it is important to recognize the vast differences that babies can present and that the attachment process, while achievable with even seriously challenged children, may require professional support and coaching. To assume that any difficulty in attachment lies on only one side is a barrier for many capable parents who are parenting a child with developmental differences that interfere with the process. It is apparent to those of us who have had this experience that little brains can be engineered before birth in ways that make relating extraordinarily challenging and produce tremendous frustration, grief and loneliness for the caregivers. These parents and babies are among the most vulnerable to relational trauma and are better served by a model that looks at both parent and child behaviors as a point of assessment and intervention.

To this end, Dr. Kathryn Barnard's Nursing Child Assessment Satellite Training (NCAST) process is invaluable. Originally developed at the University of Washington, NCAST relies on the observation of typical parent-child interactions—usually in the home—during each feeding and each teaching sequence. Once engaged in the task, parent and child behaviors are noted by the observer—usually a trained nurse—on a standardized binary scale ("yes" or "no," the behavior happened or did not happen), resulting in very clear behavioral feedback on tiny cues (like gaze aversion) that a parent might have missed or misinterpreted. This is a masterful set of tools for teaching infant states and cues and can be used to translate the behavioral language of babies into an ongoing opportunity to enhance the attachment process.

It is important that attachment classifications not feel like an indictment to foster and adoptive parents, many of whom inherit various degrees of trauma in children who have at the very least experienced separation from their biological parents. One can't say enough about the kind of commitment it takes to facilitate healthy attachment in children whose lives have been shattered. While they usually do well in spite of it all, parents of children with chronic medical conditions often face serious challenges enabling secure attachment, particularly when the child's experience is one of chronic pain in earliest development.

The good news is that if attachment challenges are caught early, especially when therapeutic support is available, even a child with disorganized attachment may become securely attached. Attunement is simply much more challenging with atypically developing children, and chronic mis-attunement is likely to require specialized help.*

The Right Brain

The question remains: how exactly does lack of secure attachment set the stage for illness, both mental and physical? According to neuropsychologist Dr. Allan Schore of UCLA, the answer lies in understanding the right brain and the effect of attachment on the coping mechanisms of this unique region. For years Schore has been the wizard of the right orbitofrontal cortex, researching, writing and speaking on the wonders of this underappreciated region of the brain. His pioneering work has integrated psychology, neurobiology and attachment theory. Schore considers the right brain—specifically the right orbitofrontal cortex—the linchpin in attachment and subsequent emotional health. He believes that failed or maladjusted attachment is the foundation of several psychiatric problems. His latest work includes redefinitions of several mental health diagnoses for the upcoming fifth edition of the *Diagnostic and Statistical Manual of Mental Health Disorders (DSM-V)* to reflect their neurobiological roots in early attachment trauma.

From the last trimester of gestation to age two, we experience the most rapid brain growth of our lifetime. This growth is not entirely symmetrical. The emotional right brain undergoes a growth spurt in the first two years of life that creates larger hemispheric volumes on the right side than on the left. The right side dominates development through the end of the child's third year after birth, when spoken language begins to flourish and stimulates growth on the left side.

So, in spite of the fact that language is primarily the domain of the left hemisphere, linguistic development actually begins in the right hemisphere with the infant's orientation to the face, voice and gestures of the caregiver. For our infant selves, learning is fundamentally emotional; cognitive learning will follow, and its success has everything to do with how

---

*For any parent in this situation, the techniques taught by "Floortime," Dr. Stanley Greenspan's curriculum for parents of developmentally challenged children, and the "Circles of Security" work by Kent Hoffman and Bert Powell (www.circlesofsecurity.org) are invaluable.

well early caregivers have facilitated emotional regulation in the baby. In the beginning, life is all about feelings and sounds, tastes and textures, images and gestures, all of which is processed and stored in implicit memory in the right hemisphere. Not surprisingly, then, it is also our mother's right hemisphere that needs to attune and respond. Gestures, smells, touch, faces and sounds—especially prosody, the lilting tones and tempos that babies inspire—are the language of the baby and the attuned parent. Although most language is processed through the later-developing left hemisphere, the early-dominant right hemisphere—the seat of our unconscious perceptions—remains central to the most critical aspects of social communication throughout our lives and the core of our relationships.[9]

The orbitofrontal cortex is particularly tuned to human facial expressions and tones of voice. It matures in the middle of the second year of life, when the average child has a vocabulary of fewer than seventy words. "The core of the self is thus nonverbal and unconscious," Schore notes, "and it lies in patterns of affect regulation."

> The first phase of linguistic development begins with a growth spurt in the right hemisphere during the last trimester of gestation and continues for five to seven months after birth. The second phase, which continues from twenty to thirty-seven months, is social and all about reciprocal interactions and emotional expression in the form of utterances or primitive sounds intended to express the baby's state and needs. Sometime around thirty months the right hemisphere forms an interactive system with the later-maturing left brain, an achievement that allows for the emergence of speech.[10]

Much faster than the left side, the right orbitofrontal cortex continues to connect incoming sensual data from the cortex with our simultaneously occurring awareness of our internal physical and emotional states. Here is the seat of our "gut" reactions, our intuitive feelings about people and situations. If you are an empathizer, that feeling of discomfort in the face of someone else's suffering is registering in this part of your brain, and it began when someone was there for you as a baby who responded sensitively to your cries and needs, comforted you, and gave you a sense of connection to another person. The quick contingent response of the caregiver—her ability to read the baby's intended message and to act on it, the warmth of her response, the availability of eye contact, soothing sounds, and gestures—all formed the basis for feelings of connectedness by the baby to the caregiver and shaped the baby's model for future relationships. These feelings were mirrored by chemical and structural correspondents being built into the infant's orbitofrontal brain. If this foundation is missing, empathy can be compromised or lost to a child—and to our world.

But that's not all. The right orbitofrontal cortex also controls and modulates the primary emotions, including, and perhaps especially, fear. The right brain is more deeply interconnected than the left hemisphere with the autonomic, limbic and arousal systems, including the vagus nerve, and it plays a stronger role than the left in the production of cortisol and in key immune, endocrine and cardiovascular functions.[11] So here in the right brain is the seat of the circuitry modulating both sympathetic and parasympathetic functions. The right orbitofrontal brain governs the autonomic nervous system. Because orbitofrontal feedback assessing the level of threat provides the calming of extreme fear states, a person with an impaired orbitofrontal system is susceptible to pathological states of dissociation when stressed. Dysregulation of this feedback loop places the person at the mercy of an amygdala-driven defensive state longer than is healthy.[12] This means that our ability to manage strong negative emotions, especially fear, is built in a part of the brain that develops most rapidly and fundamentally in our first two years, primarily through experience. The ability to calm oneself, originally provided by the caregiver, is internalized by the baby, then the child, then the adolescent, and finally the adult through the attachment relationship. Without conscious awareness, we either learn to return efficiently to a calm state following stress or we remain hypervigilant, wary, on edge, or, in extreme cases, frozen. The basic template is set on an organic level.

Dr. Gabor Maté tells a story about his own nervous system. Maté was born in Nazi-occupied Poland. When he was a tiny infant, his parents were overwhelmed, fearing not only for their own safety but also for the safety of their parents, who ultimately died in Auschwitz. Maté says that at one point his mother, concerned about his incessant crying, called the pediatrician to come to their home to see him. The doctor responded that he would certainly come, "but," he added, "I have to tell you, Mrs. Maté, that all of my Jewish babies are crying."[13] Maté, who has recently published a book on addiction, says that "according to all studies in the United States as well as Canada," the hard-core addicts he treats are, "without exception, people who have had extraordinarily difficult lives. And the commonality is childhood abuse. In other words, these people all enter life under extremely adverse circumstances. Not only did they not get what they need for healthy development, they actually got negative circumstances of neglect. . . . That's what sets up the brain biology of addiction." Maté recognizes that his own trauma was buffered by a strong attachment to a loving and competent mother, but nonetheless he was affected by circumstances over which his parents had no control. He attributes his own less debilitating forms of addiction as an adult, including shopping for the "high" and workaholism, to early trauma.[14]

If we have had a securely attached relationship, we build an autonomic system that will accelerate under stress but then will find its way back to homeostasis relatively quickly. These are the fortunate ones among us. If there has been chronic early relational trauma, our capacity for self-regulation will be affected, paving the way for mental and physical diseases. The body does in fact keep score.[15]

Inability to self-regulate is at the core of most of our world's miseries, from terrorism, war, environmental degradation and child abuse to personal relationships, diet and disease, mental, physical and behavioral problems. Because self-regulation begins with the physiological regulation of the infant by an adult and because the process is built into the baby's brain, and because it shapes behavior and health for a lifetime, attunement and secure attachment are among the most crucial skills to protect within our families. Emotional and physical regulation is inextricably part of the same process. We experience regulation and contentedness, or dysregulation and alarm inside of relationships, beginning in the womb.

As Americans, we are seduced into the illusion that "having things" bestows happiness. Anticipating the arrival of a new baby or subsequent celebrations of a child's birthday becomes an exercise for most of us in shopping for "stuff." Toys, videos, CDs, equipment and clothing are enticingly marketed to convince us that the child's safety, health and intelligence depend on the right purchase. The time, income and energy absorbed by such diversions are powerful distractions from our ability to deliver what really matters. For babies, the best toy of all—and the best investment in a child's health, safety and IQ—is the focused, receptive attention of someone who views that child as having "hung the moon." This is the single best opportunity for scaffolding lifetime health. The attachment relationship is nature's "security blanket"—our primary buffer against skewing the HPA axis and the autonomic nervous system. But many of us have already missed this protection. The question then becomes: what can we do to prevent or alleviate the impact of trauma when our "security blanket," or that of someone we love, is in shreds?

OUTTAKES

This morning Meredith and I went to a local bakery/coffee shop with two friends. We sat next to a pretty young woman with a baby girl in a car seat latched into a stroller. As we sat down at our table with our coffee, I noticed the young woman was on her cell phone. The baby sat staring wide-eyed at the young woman while she chatted for fifteen minutes nonstop, never looking at the baby. The stroller was less than six inches from my chair, so that as the young woman got off the phone, I commented on how pretty the baby was and asked her if the baby was hers. "No," she said, with no pleasure at my comment, "I'm the nanny." The baby sat quietly in the stroller for over an hour, sucking on a pacifier, continually staring at the woman. Her little feet were bare and cold. A blanket folded over the stroller handle dropped to the floor unnoticed and was not picked up until they left. The "nanny" continued several phone conversations and was finally joined by another young woman with whom she talked animatedly, her back to the baby, whom she said was ten weeks old. The car seat/stroller was elegant, the young woman fashionable and clearly from a privileged background. The baby was dressed as if for a photograph. But between this caregiver and her little charge, there was no engagement, no reciprocal eye contact, no communication between the two for more than the hour and a half that we sat within inches of them.

This scenario is repeated hundreds of thousands of times every day in communities across our nation. The particulars may vary but the bottom line does not—no connection. Most people would not identify this picture as one of neglect . . . the accessories and the setting were misleading. Few would think about such a scenario as trauma to a ten-week-old nervous system. What we observed was, at the very least, a wasted opportunity. At ten weeks, babies have an unparalleled window for attachment as a foundation for health and well-being for now and in the future. Each missed effort on the part of the caregiver, each turning away, each unrewarded effort by the baby to connect—these are tiny traumas that accumulate over time for a baby—and for our society.

CHAPTER 9

Rock-a-Bye

Therapy and Beyond

We do not see things as they are. . . . We see things as we are.

—TALMUD

LIKE A ROCK THROWN into a pool of water, toxic fear and trauma early in life generates a ripple effect. The younger the victim, the greater the ripple and the deeper the rock submerges. The deeper it submerges, the harder it is to detect the original source of the ripple. Chapters 1 through 8 have summarized the research that allows us to understand the physiological processes that connect health to chronic fear in early childhood. We now consider where a victim of early trauma can turn for comfort and healing in the great lag between scientific discovery and the reality of our lives.

SIGNS OF TRAUMA IN ADULTS

Most of us have experienced emotional trauma. Most of us recover and go on with our lives, though not without repercussions. Those of us fortunate enough to have gotten off to a protected start are the lucky ones. But even for those who began life under adverse conditions, research provides vibrant testimony to the plasticity of the human brain and confirms that we have a great capacity to heal over time.

215

Sometimes a history of trauma is apparent. More often it is not. Symptoms of trauma—anxiety, nightmares, social withdrawal, emotional numbing—reflect a narrowing of behavioral repertoires to avoid aversive internal sensations, a sort of psychological inflexibility that results in our feeling removed from being present in minute-to-minute reality. We may have flashbacks, phobias, depression, insomnia, unstable moods, aggression, delusional thoughts, substance abuse, chronic pain and disease. While a traumatic brain injury from a physical blow shows up on an MRI, emotional trauma seldom does. To make things more complicated, the memory of the trauma may be hidden, unavailable to one's conscious mind. Yet the repercussions are similar: dissociation or chronic hypervigilance, or both. The resulting dysregulation of the autonomic nervous system will set off a chain of physical aftershocks in bodily systems—digestive, respiratory, cardiovascular, immune and endocrine.

Health is about the regulation of key systems in our bodies. Healing from trauma is not just about controlling symptoms but also about promoting the balanced regulation of vital systems in our own bodies and in the developing bodies of our children. This concept is fundamental to the newer trauma therapies. Unlike traditional therapies that rely chiefly on talk, several new techniques address not just cognitive and emotional processing but physical processing as well. This is not to imply that more traditional forms of therapy (for example, cognitive behavioral therapy [CBT], which has been shown to be effective with PTSD) are not helpful for trauma or to overlook the fact that many individual practitioners of classical forms of therapy have integrated an understanding of body-centered work into their practices. But most traditional talk therapy, while grounded in an understanding of the formative role of childhood, often fails to access traumatic experiences that occurred before we could speak. Moreover, many traditionally trained therapists are uncomfortable working directly with the body. Mary Sykes-Wylie, senior editor of *Psychotherapy Networker* magazine, observes:

> We're not, to say the least, a culture at home with the body. From early on, we're taught to regard all that stuff below our heads (forgetting that our heads and their contents are part of our bodies, too) as somehow distinct from our real selves, the "I" and "me" hovering invisibly just be-

hind our eyes. For many of us, the idea that our entire experience of the world and what we make of it, the way we think and feel and act, all of what we call ourselves "happens" in the very tissues and fluids of our bodies seems bizarre, impossible, almost scandalous. We seem to think of our bodies as an odd sort of property—perhaps an interior, jungly, not particularly desirable quantity of real estate—that we inhabit uneasily, subject to the peculiarities of its climate, its inconvenient rhythms, its urgent necessities.[1]

Yet paradoxically, American society is arguably the most body-conscious society on earth. We are obsessed with how our bodies look, how they perform, how others perceive and judge us. We spend billions of dollars and as many hours trying to diet, exercise, buff, depilate, cosmeticize and surgically enhance our bodies so they resemble the media icons of perfection constantly before our eyes. Many of us are health worrywarts, scrutinizing every food label like Talmudic scholars, adhering to Olympian exercise schedules, constantly looking for suspicious moles, bumps and rashes, aches, or tingling sensations, and wondering if we ought to go in for a whole-body MRI, "just in case." As Sykes-Wylie says, "It would all be hilarious if it weren't also so sad."[2]

WHEN WORDS AREN'T ENOUGH: THERAPIES THAT "RESET" THE BODY

The good news is that a new generation of therapeutic skills are now available that are designed to treat the brain and the physical impact of trauma, including early trauma, either separately or as an adjunct to a more traditional approach like CBT. In these therapies, there is recognition of the role of implicit and unconscious memory. The techniques target the integration of physical sensations, feelings, gestures, postural changes and other physical signals associated with stored trauma. The common denominator in these approaches is a combination of the gradual recall of traumatic memories (exposure) with relaxation techniques designed to modulate dysregulated bodily systems. Francine Shapiro, creator of eye movement desensitization and reintegration (EMDR), puts it this way: "When a memory of a past event is *functionally* stored, it is in declarative or narrative memory. If it is *dysfunctionally* stored, it is in motoric memory and retains the physical sensations and high level of affect that was there at the time of the event."[3] EMDR and most of the other trauma therapies seek to move traumatic memories from motoric or procedural storage to brain areas that add understanding through rational or linear processes. This integration can be accomplished by various methods in addition to EMDR: self-regulation therapy (SRT), sensorimotor therapy, "tapping" or emotional freedom technique (EFT) and "brainspotting."

These therapies reflect ancient understandings. A glance at humanity's tribal origins across the world reveals the use of organized rituals, rhythm and dance at times of crisis and transition. Ritualized ceremonies, including dance and drumming (bilateral movement) and chanting and singing (vocalizations), are part of the fabric of all primitive cultures, particularly preceding and following transformative events in tribal life: war, marriage, birth and death. These ancient practices involving side-to-side movement, music, storytelling and a deeply shared sense of interdependence take on new significance as we review what we have learned in treating trauma.

Play Therapy

My own exposure to trauma work employing movement and abstract expression began more than a decade ago when I traveled to a little community in Pennsylvania to give a keynote talk on *Ghosts from the Nursery*. Aware of my interest in early childhood, organizers of the conference took me to a preschool program they had developed based on the Artists in the Schools Program (in which local artists work with schoolchildren). In this little town, they had decided to apply the program to early childhood, targeting several preschools and child care facilities that served low-income children. We drove to a little theater that had once been a playhouse—a quaint and lovely old structure that had a formally draped stage and wooden theater seats and housed a multilevel arts program, including this one for preschoolers. Combining playacting, storytelling, drumming and exposure to varied and unusual musical instruments like the didgeridoo from Australia, congas from Africa, and flutes and horns from across the world, two young male teachers, who were clearly having a great time in the dissonant but enthusiastic din, directed the children as they played the instruments in rhythmic songs.

Then the children—the majority of whom I learned were living under very challenging circumstances, including poverty, homelessness, domestic violence and foster care—proceeded to the stage for a storytelling session with one of the teachers. I watched alone from the front row while the children sat themselves in a circle around him. In vibrant tones and with animated expression, the teacher began to read a story, setting a scene about a little monkey in the jungle with tall trees and a forest bursting with tropical fruits and flowers. There were minor characters that called for voices and action, such as birds and a gazelle. But gradually he directed all of the children to act out the roles of the central characters: a little monkey and a lion, the monkey's fierce challenger for life in the jungle. Eventually, as the children plucked bananas from tall trees, ran and roared and scampered and stomped and hooted, everyone was a little monkey who, after lots of narrow escapes, outsmarted and rose victorious over the lion!

Unprepared for the purpose or passion of the session, I found myself in tears as I watched the profound impact of the exercise on the children. The reenactment of the fear that many of the children had absorbed and their expressions of rage and terror and cunning and then victory over their predator took me by surprise in their vehemence. All the elements of dynamic play therapy had just taken place—connection with a trusted adult through play and daily interaction, relaxation, bilateral stimulation of little brains and a new narrative of triumph. While for most children who experience early trauma a single session would probably be only the beginning, it was a powerful step toward healing. This was a therapeutic approach that would have a huge impact on learning, one that could easily be incorporated into educational settings for children with similar needs.

Play therapy is a powerful tool for healing trauma, particularly when it allows for a form of re-exposure and simultaneously integrates movement and bilateral stimulation designed to release the mobilizations frozen in trauma. Compared to adults, children tend to heal rapidly. Utilizing various types of toys—stuffed animals, dolls, trucks, cars, blocks, sand, clay—therapists can facilitate the expression and resolution of trauma. Often through play alone, young children will act out their memories in metaphoric and third-person terms. The use of soothing materials, like sand and water, and dance and music adds elements of relaxation and sensual stimulation that both complement the "story" and facilitate the integration of procedural with declarative memory. The sooner children can be placed in a healing environment, the better the outcome.[4]

Since my exposure to the Pennsylvania children in the little theater, I have seen several outstanding arts programs for children that have very similar goals. One type of activity that is not necessarily designed for therapy—and thus escapes the stigma of such—but is exquisitely therapeutic is dance programs for inner-city teens, some of whom have absorbed trauma in many forms, including poverty, racism and the gamut of "adverse childhood experiences." A dazzling example in Portland, Oregon, is Urban Arts. The dance company and school is run by exceptional professionals who help young dancers create their own choreography about their life experiences, utilizing the best of current dance technique and music—from rap to techno to tribal African—along with original poetry, staging, sound and costuming that rivals the best productions anywhere. The dancers put in long hours, often for years, dramatizing through dance their experiences of love and loss and their emergence through these experiences.

The arts are a powerful therapeutic tool, particularly dance, which involves consistent bilateral stimulation and vigorous rhythmic activity. Clearly "therapy" doesn't always take place in an office! Work with animals, guided experiences in the wilderness, and constructively coached sports training also facilitate bilateral stimulation of the body and brain and have their place in mitigating the toll of trauma in a human life.

Family Systems Therapy for Children

For most traumatized children, the very best therapy is a loving family. When a child is acting out symptoms through nightmares or aggression, therapeutic play and other forms of individual therapy may be essential. But whenever feasible, it is critical to involve parents—and sometimes other key adults, such as grandparents or older siblings—in the therapeutic process. Not only will engagement of the entire family often resolve the problem more quickly, but it may also avoid singling out the already traumatized child as the designated patient, which might add to an already traumatized system. Whenever appropriate, working with the whole family and giving parents the tools to be the child's therapists on an ongoing basis extends the therapeutic milieu. Systems work has the added benefit of offering healing to the entire family when trauma has affected them all or the child's symptoms are affecting everyone.

Eye Movement Desensitization and Reintegration

In 1987 Dr. Francine Shapiro was troubled by her own disturbing thoughts when, as legend has it, she decided to go for a walk. As she walked her thoughts were playing out in her mind. She noticed that as she strolled and thought about her problem, her eyes were darting back and forth in a specific way, "a very rapid ballistic, flicking movement." After her walk, Shapiro noticed that her disturbing thoughts had faded: when she refocused on them, they didn't have the same emotional charge. Shapiro, a psychologist, began experimenting, then researching. The result was eye movement desensitization and reintegration (EMDR), one of several relatively recent therapies that use the body as the primary entry point in processing trauma.

EMDR therapy consists in part of bilateral stimulation of the senses: sight, sound and touch. We do not know why trauma can be healed by rhythmic touch on alternating arms or legs, or oscillating tones from one ear to the other, or a metronome-like visual cue (usually the therapist's fingers) moving from side to side. One theory is that it somehow facilitates movement of the memory from one side of the brain to the other, from the nonverbal, motoric and implicit memory to the verbal and analytical explicit memory, allowing integration and relaxation of the fight-flight-freeze tension that accompanied its storage. EMDR and many of the sensory processing therapies seem "magical" in their effects. In applications of EMDR to young children, for example, the diffusion of the symptoms may occur in less than a handful of sessions, sometimes in only one or two. There is a lag in the science between the observable impact of these interventions and a biological explanation of how they work. In spite of initial resistance from the traditional therapeutic community, there have now been thirteen controlled studies of EMDR. The most recent of these indicate that 84 to 90 percent of single-trauma victims are relieved of PTSD after only three ninety-minute sessions. Shapiro says: "With EMDR we see clients start at a high level of affect and physical sensations and after treatment, that is no longer there, and learning has taken place."[5] Most people experience relief within one to twelve sessions.

Self-Regulation Therapy and Sensorimotor Therapy

In addition to EMDR, several other therapies target self-regulation and sensorimotor techniques. In each, the individual is encouraged to direct their attention to their own internal feelings, sensations and behaviors that arise as they revisit the traumatic event verbally with the therapist. The goal is to attain or regain the ability to modulate emotional and behavioral responses and the ability to self-soothe.

In self-regulation therapy, the therapist is trained to discern subtle signs of the client's sympathetic arousal or parasympathetic down-regulation as he or she recounts a traumatic experience. The therapist uses his or her attunement to these signals to teach the client to recognize and modulate felt physical and emotional responses as these surface during the recounting. Cognitive focusing tools—such as breathing, various forms of alerting clients to their own bodies and directing attention to physical aspects of the immediate environment—are used to have clients experience a horrific memory while simultaneously experiencing and recording a more modulated or benign feeling. The result is a gradual defusing of the initial experience and the recording of that difference in the brain.

The challenge for the practitioner is to calibrate exposure to the memory so that it does not overwhelm the client. The therapist also helps the individual identify resources within their own skills and experience—friends, family, memories, aspects of nature and place, humor, passions, conscious attributes or strengths, as well as places within their own bodies that stabilize their internal state. From the therapist the client learns a gradual process of self-regulation and the modulation of overwhelming and formerly out-of-control emotional and physical processes, thereby regaining a sense of control over what had been unmanageable symptoms. The therapist's role resembles that of a skilled caregiver in a securely attached relationship: gently noticing physical cues that indicate discomfort, putting that into words, redirecting attention to what is occurring now, and reminding the individual of his or her own capacity to gentle the experience (like breathing).

Through the language of the lower brain—breathing, heart rate, temperature, muscle tension, visual cues, reported sounds or smells or tastes, and involuntary movements like twitches and postural changes—SRT provides a bridge to procedural memory. Trauma memories are stored along with the sensations and feelings experienced at the time. For reintegration to occur, unresolved sensory motor reactions must be experienced, especially defensive reactions, such as running, hitting or screaming, that the person would have completed if he or she had been able. Inhibitory thoughts from the cortical brain keep trauma victims captive in fight-or-flight, compromising their health. Once they are accessed through behavioral correlates, bypassing the inhibiting controls of the rational brain, defensive maneuvers are completed and held energy is released. A titrated unwinding of the freeze response allows trauma to be discharged.

Many body-focused therapists view the creation of a cognitive framework for understanding what this means to an individual (meaning-making) as a second phase. SRT is a particularly good therapeutic match for treating diseases connected with autonomic dysregulation, such as asthma, allergies, autoimmune disorders, fibromyalgia, whiplash syndrome and chronic pain.[6]

Sensorimotor therapy is similar but not identical to SRT. The primary difference is that tracking physical sensation is often an end in itself in SRT. Sensorimotor therapists feel strongly that physical tracking, the regulation of reactions and the expression of blocked responses are not enough. In sensorimotor work, clients are taught to distinguish trauma sensations from emotions. Once they have learned to self-regulate arousal through sensorimotor processing, the focus shifts to include cognitive and emotional processing and to address relational dynamics. Sensorimotor therapy is particularly useful in the treatment of relational trauma as well as shock. Sensorimotor therapists report strong success with PTSD clients in reducing nightmares, panic attacks and aggressive outbursts.[7]

Brainspotting

David Grand, the originator of "brainspotting," defines a "brainspot" as an eye position paired with externally observable reflexive responses that are associated with the activation of traumatic memories. Brainspotting therapists are highly attuned to signs of arousal as indicated by subtle, often minute behaviors—an eye blink or pupil dilation, a sniff, a swallow, a yawn, a foot or hand movement—that signal that a brainspot has been located and activated. The brainspot can then be accessed and worked with by holding the client's eye position while the client focuses on the somatosensory experience of the underlying issue that is causing distress. Holding of the eye position while focusing on internally felt physical sensations is believed to stimulate a deep healing process within the brain. The same technique can be used to access and integrate internal resources. The brainspotting process is often enhanced by the use of bilateral sound to support the integrative process.[8]

Emotional Freedom Technique or "Tapping" Therapy

Emotional freedom technique (EFT) is said to be an emotional version of acupuncture. For individuals working with an EFT-certificated therapist, the technique consists of imagining the troublesome issue (or experience or sensation or feeling), rating their level of discomfort on a ten-point scale, then tapping with their fingertips on specific meridians, as taught by the therapist. The activity is thought to facilitate the integration of right- and left-brain activity. The approach is gentle, involves no pain, and is fast, providing relief in weeks or months and sometimes instantaneously. Practitioners claim that painful sensations or feelings simply disappear. The technique does not require a professional therapist. In fact, the originator of EFT, Gary Craig, is an engineer, not a therapist. He makes the training materials widely available to anyone without charge.[9]

LAYING THE GROUNDWORK

Finding a therapist or deciding what form of therapy to seek can seem overwhelming, especially for anyone who is already physically and emotionally struggling. It may help to know that there is little to suggest that one approach is significantly better than the rest. A University of Wisconsin meta-analysis of the relative efficacy of various psychotherapies for treating PTSD yielded no differences between them and concluded that, while there was strong evidence of the efficacy of therapy compared with no treatment, the bonafide psychotherapies (in this case including but not limited to cognitive behavioral therapy, EMDR and many others) produce equivalent benefits for PTSD patients.[10]

Potential clients curious about how to evaluate therapy will find some answers in another interesting meta-analysis that looks at predictors of positive outcomes in psychotherapy. Reported by Mark Hubble, Barry Duncan and Scott Miller in *The Heart and Soul of Change*, this analysis found that more than 40 percent of positive outcomes appear to be attributable to something that happened outside of therapy. Looking closely at the remaining 60 percent, the report concludes that the best single predictor of positive outcomes in therapy is the development and maintenance of a trusted relationship with the therapist, combined with the client's perception of hope and positive expectancy. Key is one's sense that the therapist sees and trusts the client's abilities, as well as his or her own, and believes in the client's intrinsic capacity to heal.[11]

For those of us who have been through trauma, the first step is to establish safety—real and perceived. If we are in real danger in our daily lives, we have to find a safe place to be. Similarly, if we perceive even subtle threat within the therapeutic relationship, progress will be nonexistent or inhibited. Low levels of fear will continue to stimulate the sympathetic nervous response, undercutting the ability to calm the amygdala, which is central to all that follows. Right brain to right brain, gesture by gesture, the nervous system of the therapist attunes to and leads that of the client, noticing somatic cues and tiny nuances that indicate autonomic imbalances. Guided relaxation or hypnosis may be useful tools to access the traumatic experience. In the hands of a competent therapist, aspects of the trauma are now experienced differently as they unfold. Coupled with exercises designed to relax the sympathetic system, pieces of the recalled trauma are experienced with increasingly gentled and modulated physical responses until the recalled memories lose their capacity to elicit the extreme sympathetic reactions of the previously recycling trauma. Exposure is paired with relaxation and increasing self-regulation until pathological reactions evaporate. The process can take a few sessions, a few months or even years. The work is simple—but not easy.

The topic of what constitutes sound and appropriate "therapy" is hotly debated. Matching yourself to the right approach is important, but more important is finding the right therapist. Given adequate credentials, education and experience, the most important criterion is the sense of connection and trust you feel in the therapist's presence. We get sick inside of relationships, but we also build health inside of relationships. Attunement is fundamental to healing. Because traumatized individuals and those who have suffered "adverse childhood experiences" are particularly likely to be vigilant about how they are perceived, this relationship is key. Emotional distance on the part of the professional, the traditionally ingrained detachment and objectification of the relationship with the patient, while helpful for a surgeon, is contraindicated for this kind of therapy. In describing healing relationships, Dr. Daniel Siegel says:

> Connections between minds therefore involve a dyadic form of resonance in which energy and information are free to flow across two brains. When such a process is in full activation, the vital feeling of connection is exhilarating. When interpersonal communication is "fully engaged"— when the joining of minds is in full force—there is an overwhelming sense of immediacy, clarity and authenticity. It is in these heightened moments of engagement, these dyadic states of resonance, that one can appreciate the power of relationships to heal the mind.[12]

Psychotherapy has historically been geared toward psychological or emotional goals without recognition of the physical underpinnings of state regulation. Awareness of the body and the physiological signs of internal function adds a new language to that of traditional therapy. While each major school of therapy—dynamic, behavioral, cognitive, experiential— is effective for particular issues, the recognition of physiological cues and close physical attunement has the potential for treating early trauma more directly and often more quickly than traditional approaches. Dr. Allan Schore believes that attachment theory is essentially a regulation theory: a framework that explains and supports the relational roots of health and disease. Therapy is an opportunity to reinstate or create modulated self-regulation of vital psychobiological processes, a form of reparenting—restoring homeostasis and equilibrium.[13] It appears that when trauma is the problem, attachment is a key piece of the solution.

TELLING THE STORY: THE PERSONAL NARRATIVE

Once the most problematic symptoms of trauma are reduced and individuals have established relative confidence in their ability to self-regulate flooding emotions or sensations, they are ready for a further step in integrating fragmented pieces of a previously traumatizing experience. An integrated chronological story that encompasses the experience in words and is shared with another person—someone who bears witness to and validates the struggle—is useful where trauma has been chronic, early and relational.

The narrative process can begin in many ways, including photographs, memorabilia or stories gleaned from family members or friends. Some clients show up at the therapist's office with a large box of albums or loose photographs. The therapist gently facilitates the story, listening while being aware and directing the client's awareness to his or her own "lifeline," the intrinsic strengths, skills, attitudes and decisions he or she used to survive and prevail. Sharing observations of the individual's resourcefulness, persistence, creativity, wiliness, adaptability and so on, is powerful since many if not most of us who have suffered trauma tend to feel marked, embarrassed or even "crazy" because of the contrast between our lives and what we believe is "normal."

In this therapeutic role, one cannot help but be moved by the courage and strength of the human spirit. Of all the responses we got after publishing *Ghosts from the Nursery*, the one we heard most was gratitude from readers who for the first time felt that they weren't crazy because they now understood their symptoms in terms of the physical impact of abnormal experiences on their brains. A simple explanation of brain science, basic anatomy and the hormonal processes that characterize trauma is extraordinarily helpful to many clients at this stage. This is an intimate and intense but inspiring journey, and both patient and therapist are changed along the way.

SELF-DIRECTED HEALING:
EXERCISE, RELAXATION AND QUIETING THE MIND

There have been countless studies of the role of exercise in protecting health and preventing many chronic diseases, including cardiovascular disease, osteoporosis, diabetes, osteoarthritis, weight gain and mental health problems, particularly depression and anxiety. For all of us who are able, it is important to move our bodies every day, and aerobically at least three times a week, working up gradually to a minimum of thirty to forty-five minutes of exercise at a time. What kind of exercise we engage in is less important than the fact that it is aerobic: aerobic exercise is crucial to the circulation of natural serotonin and the reduction of norepinephrine. When trauma is the issue, movement becomes even more critical, though a traditional Western approach to exercise alone may not be enough.

The good news is that there are several proven paths to quieting the autonomic system and down-regulating a hypervigilant amygdala. Yoga and other Eastern practices that re-establish homeostatic balance are invaluable for many. The best way to know whether or not it works for you is to try it out. You can assess the usefulness of such a practice by measuring your heart rate, heart rate variability, blood pressure and (slightly more clinical but available) levels of norepinephrine, cortisol and serotonin. Yoga also brings about obvious and immediate changes in breathing, muscular tension and feelings of calmness and comfort with one's world. Researchers at Boston University School of Medicine have found that practicing yoga elevates GABA, the brain's primary inhibitory neurotransmitter, which is associated with depression and anxiety disorders. Since raising GABA levels is a primary goal of prescriptive medications, yoga offers major benefits, to say nothing of increased flexibility and a better back and posture. Using MRI imaging, researchers found that after only one hour of yoga, participants had a 27 percent increase in GABA compared to a matched group who read for an hour.[14]

In addition to yoga, several other Eastern practices are showing evidence of effectiveness in stress reduction, although much of that evidence is still subjective. Both tai chi and qigong ("chee-gong") show promise, particularly for releasing the held tension of frozen trauma. Both practices focus the mind on bilateral motoric movements and breathing that help accomplish the discharge of blocked fight-or-flight energy. Although there is no research to date on the impact of martial arts in the treatment of trauma, there is some indication that such practices have a positive impact on the immune system. Tai chi chih, a nonmartial version of tai chi, is characterized by slow movement and combines meditation, relaxation and some aerobics. A twenty-five-week study of 112 adults ranging in age from fifty-nine to eighty-six showed that the regular practice of tai chi chih significantly boosted their immunity against the virus that leads to shingles. The study divided the individuals into two groups. Half took tai chi chih classes three times a week for sixteen weeks, while the other half only attended health education classes, where they received advice on stress management, diet and sleep habits. After sixteen weeks, both groups received a dose of the shingles vaccine Varivax. At the end of the twenty-five-week period, the tai chi chih group had double the immunity levels of the health education group and also reported significant improvements in their physical and mental health and reduction of bodily pain.[15]

Meditation is another path to gentling the amygdala. In 2004 researchers at the University of Wisconsin found that meditation actually alters the physical circuitry of the brain by increasing gamma brain waves.[16] Gamma brain waves are associated with increased levels of perception and awareness, higher levels of intelligence, energy and focus, better memory, and positive moods and thoughts. Forms of hypnosis and self-hypnosis and even simply adequate sleep are also known to increase overall gamma wave levels, enhancing our sense of positive emotionality and self-control—a quieting and health-protective opportunity that we can accomplish with a relatively small investment of time or money.[17]

Meditation has become such a popular approach to stress management that it is being used in some corporate and organizational settings to support employee health and productiveness. The menu of choices usually includes traditional Buddhist practices, transcendental meditation, a simple tried-and-true approach designed by Dr. Herbert Benson of Harvard Medical School called the "relaxation response," and many creative hybrids. Emerging in popularity, particularly for stress management, is "mindfulness," a form of meditation pioneered by Jon Kabat-Zinn in the early 1990s as a method of dealing with stress and chronic pain. Mindfulness has also been incorporated into several forms of therapy, including mindfulness-based cognitive therapy for the prevention of depression relapse, dialectical behavior therapy for borderline personality disorder, and mindfulness-based relapse prevention for addiction treatment.[18]

Kabat-Zinn defines mindfulness as "the awareness that emerges through paying attention on purpose, in the present moment, and non-judgmentally to the unfolding of experience moment to moment."[19] Although many forms of meditation are grounded in deeply spiritual and religious paradigms, mindfulness meditation uses progressive, systematic exercises to tune into awareness of one's internal functions and awareness of one's feelings and thoughts, without judgment. Mindfulness teaches acceptance of whatever is present, being aware in the moment, and allowing oneself to feel whatever physical sensations accompany that experience—observing thoughts without attachment to them. This deliberate shift in focus from the "what" to the "how," from content to process, enables one to observe what occurs as it happens internally. Daily practice facilitates awareness of how thoughts and feelings surface and dissolve.[20] There is extensive documentation that mindfulness constructively alters brain functions in individuals who regularly engage in its practice.[21] Through systematic training, practitioners learn new ways to accept and come to peace with worry, with repetitive and unproductive thought patterns and with unruly emotions.[22]

There is no stronger advocate for the value of mindfulness in healing trauma than Dr. Daniel Siegel, who calls for the cultural establishment of a regular "time-in." Siegel has joined together with a handful of colleagues to develop an interdisciplinary view of the mind and mental health called "interpersonal neurobiology." This construct draws on a broad spectrum of scientific frameworks—especially neurobiology and attachment—and weaves them together with expressive arts, Eastern wisdom and contemplative practices. The goal is to "harness the social circuitry of the brain." Siegel says that the regular practice of time-in creates "a documented improvement in immune function, an inner sense of wellbeing and an increase in our capacity for rewarding interpersonal relationships."[23]

Relaxing overvigilant nervous, cardiovascular and digestive systems, as well as fear-driven and impulsive behaviors, has also been the driving force behind the growing interest in both biofeedback and neurofeedback. Biofeedback enables one to learn how to gain control over bodily functions that usually occur without conscious awareness. Sensors are placed on the body to measure a given function that presents problems for the individual. Heart rate, blood pressure, muscle tension, respiration, body temperature and perspiration levels are measured and translated into a visual or audial readout. Patients learn to interpret this instantaneous feedback and immediately see how associated thoughts, feelings and images influence physiological responses, for better or worse. By monitoring the graphic relationship between mind and body, we can learn to use our own thoughts and images as cues to relax. Once learned, biofeedback equipment is no longer necessary; we have a personally tailored, broadly applicable tool to control problem symptoms.

Neurofeedback is a form of biofeedback that tracks brain waves rather than blood flow and muscular tension. Brain waves are electrical currents produced when brain cells communicate with each other and other systems in the body; they are measured by EEG through electrodes attached to the surface of the scalp. Cells firing at varying speeds produce different types of brain waves. Faster brain waves are associated with thinking, while slower ones are associated with relaxation. The first step in neurofeedback is assessment to determine the most beneficial pattern of training. The computer is adjusted to create a pleasant musical tone, and as one begins to generate the desirable waves this tone literally guides one to a calmer state. We can learn how to descend into deep relaxation while remaining quietly alert (a state that, incidentally, reflects the state of serenity experienced in the secure attachment relationship). Clients have reported a sense of empowerment as they regain capacities previously lost to trauma: feelings of calm and confidence, a good night's sleep and a sense of general well-being.

Neurofeedback is also used to treat ADD/ADHD, to reduce impulsive and violent behavioral patterns, and to alter the intrusive impact of the flashbacks and explosive, sometimes aggressive behaviors that accompany PTSD. Pioneered by the work of Dr. Eugene Peniston at the Veterans Administration Medical Center in Fort Lyons, Colorado, and Dr. Carol Manchester in Cincinnati, Ohio, neurofeedback has proven enormously helpful for veterans seeking help with serious, persistent anxiety, depression and PTSD.

MEDICATION

Without a quality therapeutic relationship, drugs are generally less effective than in combination with therapy. Nevertheless, several medications have proven effective in the treatment of trauma and its consequences. The selective serotonin reuptake inhibitors (SSRIs) are often used to treat symptoms of anxiety or depression. One of these, Welbutrin, is additionally effective for alcohol withdrawal. Psychiatrists have a growing arsenal of prescriptions for treating anxiety, depression, addiction, agitation, aggressive outbursts, insomnia and panic. An interesting discovery is that nadalol, a beta blocker originally used for treating hypertension, can be used to block the memory of traumatic experiences.

What may work best for you? Finding a therapeutic path that works for any one of us is all about trying options on for size, and there is probably no one right answer. Esther Sternberg, director of the Integrative Neural Immune Program at the National Institute of Mental Health, is internationally recognized for her discoveries in the interactions between the central nervous system and the immune system and the connections between the stress response and both arthritis and depression. She is also the author of an insightful book on these connections, *The Balance Within: The Science Connecting Health and Emotions.*[24] In an interview at the University of Minnesota following a lecture there, Sternberg said:

> There isn't one thing to do for all people and there isn't one thing to do at all times in your life. So meditating may be good at one time in your life and exercise at another. It's a constantly fluid and changing thing. . . . A healthy diet is always important. Social support is always important.

Some degree of exercise is always important . . . but if you have some reason you can't exercise, a little is better than none. . . . Developing a kind of cafeteria plan of interventions that are comfortable for you to do, that you look forward to doing, not that you have to force yourself to do. . . . The body doesn't work like a fast-food restaurant. The body takes time to heal. Whatever illness you have, it takes time. And what these kind of mind-body interventions can do is just lay the groundwork to help the body to heal.[25]

The encouraging news is that there are many new routes available for trauma therapy that are designed particularly to consider and treat the biology of trauma. But in reality, the majority of us who have experienced early trauma will for varying reasons never formally work with a therapist. Most will pursue healing through trial and error inside of relationships along the road of life. And many will heal through friendships, through prayer and spiritual practices, through helping others, through insight and wisdom learned in their journey. Perhaps the best advice of all comes from the Dalai Lama:

From the very core of our being, we desire contentment. In my own limited experience I have found that the more we care for the happiness of others, the greater is our own sense of well-being. Cultivating a close, warmhearted feeling for others automatically puts the mind at ease. It helps remove whatever fears or insecurities we may have and gives us the strength to cope with any obstacles we encounter. It is the principal source of success in life.[26]

CHAPTER 10

It's a Small World After All

The real crisis . . . is coming. It is more relentless and more powerful than the floodwaters . . . more destructive than the 150 mile an hour winds. . . . It will destroy a part of our country that is much more valuable than all of the buildings, pipelines, casinos, bridges, and roads. . . . Over our lifetime, this crisis will cost our society billions upon billions of dollars. And the echoes of the coming crisis will haunt the next generation. The crisis is foreseeable. And, much of its destructive impact is preventable. Yet our society may not have the wisdom to see that the real crisis . . . is the hundreds of thousands of ravaged, displaced and traumatized children. And our society may not have the will to prevent this crisis. We understand broken buildings; we do not understand broken children.

—BRUCE PERRY, MD,
Child Trauma Academy,
commenting on the effects of the 2005
New Orleans flood and Hurricane Katrina

SCARED SICK IN A NUTSHELL

We are an amazing species, embodying potential that exceeds our wildest imaginings. Key to our success or failure is the human brain—plastic, resilient, powerful . . . and vulnerable. We don't have to be survivors of trauma to be traumatized. All we have to do is to witness trauma. Most of us are bathed in that witnessing. Trauma generates polarities, which fracture marriages, families, corporations, communities, politics. Trauma and its effects have become ubiquitous. Maternal gestational stress, insensitive child care, divorce, addiction, mental illness and domestic violence have turned many households into war zones, especially for the children involved. PTSD is prevalent in grade school and high school classrooms. Contributing greatly to soaring rates of stress-related disease—emotional, behavioral and physical—is pervasive ignorance about babies and young children, especially as embodied in the common belief that they will "get over it" or "won't remember" and the lack of recognition of symptoms of shock and trauma in infancy and toddlerhood.

Fear experienced early and chronically triggers disease by dysregulating the HPA axis, activating the vagus nerve, and catalyzing epigenetic mechanisms that facilitate the expression of genetic disease. How this plays out for each of us depends on the balance between protective and risk factors throughout our development. Strong positive emotional connections in early development that balance the HPA axis insulate against the impact of later stress and trauma. And conversely, any diminishment of the attachment relationship between newborns and their caregivers renders the children more susceptible.

New understanding of the impact of stress on the nervous system and the HPA axis, discoveries in epigenetics, and the findings of the ACE Study all have huge implications for our daily lives and the systems we create to support our health. The following is a brief summary.

THE MIND-BODY REUNION

When it comes to our overall health, mind and body are inseparable. What happens to us emotionally resonates physically. Ancient wisdom as far back as Aristotle recognized that "the whole is greater than the sum of its parts." But this understanding has been lost in Western medicine, which allows specialization to define patients by their pathologies. We become cancer patients, heart patients, diabetes patients. When diagnosed with a major disease, we enter the bowels of the specialty and may never resurface as a whole—let alone otherwise healthy—person.

Treatment protocols are typically designed with little or no awareness of comorbid conditions that most of us have at least by middle age—for example, the woman being treated for breast cancer who is also being treated for high blood pressure. Patients referred to more than one specialist often find themselves with contradictory protocols; for example, tamoxifen for breast cancer contributes to weight gain, not good for the hypertensive. While family practitioners or internists or general practitioners may be willing to choreograph this dance of competing specialties, it is a rare doctor who is trained and comfortable—let alone has the time—to integrate the role of relationships, stress and life history into the symptoms they are seeing. Reintegrating mind and body and viewing patients holistically is essential to improving health outcomes in Western medicine.

Dr. Vincent Felitti speaks around the world on the importance of in-depth personal (familial, relational) histories that reach into earliest childhood, including anything known about a patient's gestation, regardless of the presenting issue. He is critical of the overreliance on physical examinations and lab work for diagnoses and calls for a system in which patients would fill out detailed questionnaires online that they would give to their physicians, both to improve diagnostics and to facilitate the building of a dynamic trust-based relationship between patient and doctor, a factor in health and healing all by itself. Physicians need training to learn the relational skills necessary to interview patients, discuss and integrate the information gleaned from the questionnaires into their practices, and stay on top of resources that may be supportive. The physician's knowledge of the nervous system and the HPA axis can be a powerful tool when used to help patients to understand their symptoms or disease as a malfunction or overuse of an originally protective defense by the body and essentially a normal response to abnormal circumstances.

Once trauma symptoms are redefined as the body's normal efforts to protect against what was once a threat, the stigma of pathology can be lifted and the patient has a much deeper understanding of the core mechanisms of health and disease. The role of human connections, particularly the immediate relationship between the patient and the physician, is not to be underestimated. Within this relationship lies the opportunity for trust and healing or for fear and distance. The traditional patriarchal model of physician ownership and control of pertinent medical information, communicated to patients with benevolent condescension, is antithetical to the patient's sense of ownership of his or her own health and healing.

Dr. Gabor Maté runs North America's only medically supervised site for addicts reliant on injected drugs in Vancouver, Canada. When it comes to reaching more deeply into the mind-body understanding in medicine, Maté says:

> When somebody comes in with a first episode of rheumatoid arthritis or multiple sclerosis, or even a diagnosis of cancer, it's not enough to give them pills. It's not enough to give them radiation or offer them surgery. They should also be talked to and invited and encouraged to investigate how they live their lives and how they stress themselves, because I can tell you from personal experience and observation that people who do that, who take a broader approach to their own health—they actually do a lot better. And I know people who have survived supposedly terminal diagnoses simply because they have taken their own mind-body unity, and I would say spiritual unity as well, seriously. And they've gone beyond a narrow medical model of treatment. . . . I'm not here to disparage the value of the medical approach in which I was trained. I'm just saying that it's hopelessly narrow, and it leaves many people without appropriate treatment and appropriate support.[1]

There are many therapists who see the unrecognized role of emotional trauma at the root of many if not most of the pathologies outlined in the *Diagnostic and Statistical Manual of Mental Disorders* (*DSM-IV*), the diagnostic manual that defines all forms of mental ill health. The late Dr. Stanley Greenspan demonstrated that early interventions with families that increase communication and prevent emotional trauma for the child can minimize, and in some cases totally repress expression of genetic mental illness. Emotional illness results from physiological alterations in the brain that are triggered and exacerbated by abnormal conditions and experiences. Some of these conditions appear to result from altered genes. Some, including the introduction of toxins and suspected viral infections, may begin in the womb. Some are inflicted on children through toxic stress or abuse and neglect during development.

Over time we hope that the societal view of the mentally ill and of chronically addicted people as stemming from character-based or spiritual shortcomings will gradually shift to reflect the science. Good people who get a bad disease often have a sense of personal failure—or a sense of having done something wrong or of being punished. This is particularly true when it comes to mental illness. Graphically reflecting this belief is our attitude toward homeless people, a population riddled with mental illness and addiction. We turn away, thinking, *Just get a job!* Or we assume that the homeless have a history of irresponsible choices and moral decay. Most of them, however, are the castoff remnants of serious trauma, a hard mirror for our society to confront. Increasing knowledge of neurobiology casts such notions about the homeless—like the once-common perception of mental illness as demonic possession—into the category of superstition.

This is not to argue that individuals should not be held accountable for crazy or addiction-driven behaviors and the excruciating toll they take on others. We are each accountable for own behavior. But if we understand that we are dealing with organic issues and the impact of destructive experiences on the nervous systems of the mentally ill, the severely addicted and the chronically homeless among us, we might create systems that successfully treat and, more importantly, prevent the destruction that these individuals inevitably wreak on a society that has turned its back on the root problems. Currently we pay dearly, in both human and economic terms, when we simply punish or contain the consequences of preventable trauma. Even the most sympathetic of these cases, the soldiers returning with PTSD from Iraq and Afghanistan, reflect our willingness as a society to sacrifice individuals to predictable trauma without recognizing the inevitable impact on their physical health and the financial and emotional consequences to their families—and to our nation.

America's tolerance for child abuse is another example. Child abuse is an epidemic, raging virtually unseen and unabated in our country. There were 772,000 confirmed cases of child abuse or neglect in 2008 (the most recent year for which we have confirmed data), and 1,720 of those children died. Nearly 80 percent of the fatalities from abuse and neglect were children under the age of four. Nearly 80 percent of the perpetrators were parents, and another 6.6 percent were related to the victims.[2] As bad as this number is, the actual reality is far worse. Research shows that the true number of victims—a figure that would include those never reported to authorities—may be three times as high as the official count, or over 2.3 million children abused or neglected every year. This number equals more than 80 percent of the entire population of our nation's third-largest city, Chicago.[3] The toll of child abuse and neglect in 2008 was nearly forty-one times the number of people infected by AIDS in the same year (56,000), and nearly three times the number of people who had their first heart attack (785,000).[4] But with the exception of particularly horrific cases that appear from time to time in the morning headlines and on the evening news, child abuse and neglect continue, with less public outcry—and fewer protections—than for abused animals.

There is little question but that health care costs will continue to soar as long as we sidestep this issue. Rather than waiting for little battered bodies to emerge affected by predictable risk factors that were recognizable by maternity nurses, obstetricians and pediatricians, preventive programs that begin prenatally and continue at least through the first years could save countless lives and billions of dollars. Our current approach is to wait until the same children require expensive remedial services, including child welfare, juvenile justice and prison. A handful of programs have been evaluated extensively to show strong positive outcomes in preventing child abuse and neglect (see appendix E).

THE WAR ON DRUGS

The ACE data show that the risks of disease in adulthood are magnified by the use of substances during adolescence. But when it comes to addiction, early emotional trauma is the elephant in the room, although the linkage between the two is only just beginning to permeate public awareness. As the treatment field for addiction moves forward, it is crucial that discernment and intervention in trauma as a root issue be integrated into protocols.

ALL THE KING'S HORSES

Another implication of the research is the need for increased collaboration and integration between health professionals. In treating illness, medical practices can inadvertently exacerbate disease by triggering trauma. It is highly stressful to hear a life-threatening diagnosis, to undergo surgery, or to suddenly face a radical procedure like a biopsy, MRI or radiation. Such procedures, while life-saving and necessary, are also often terrifying. Mitigating fear in medical protocols would save money and sometimes lives—to say nothing of human suffering. Medical training should include more than a glancing knowledge of why and how to minimize emotional trauma. Positive outcomes are more likely where there is easy and regular access to mental health professionals, emotional and educational support groups, and alternative relaxation skills like meditation and yoga.

As it stands, regular visits with a mental health therapist are discouraged by most health insurers and seldom paid for without a pathological mental health diagnosis that can be used—regardless of the Health Insurance Portability and Accountability Act (HIPAA)—to terminate health insurance and deny bank loans and sometimes employment. "Mental" health care is still routinely viewed as stigmatizing. What if mental health was recognized as a fundamental element of preventive care for families and made universally available, even integrated with "physical" health care? At the very least, mental health parity is critical. Fiscally focused architects of health care reform are beginning to realize that we could significantly lower health care costs by rewarding healthy behaviors that prevent disease, such as following a good diet and exercising. One strategy being tried is reducing premiums for people who exercise, maintain a healthy weight, don't smoke and get regular checkups. This approach is similar to drivers paying lower premiums for auto insurance when they invest in a defensive driving class.

AN OUNCE OF PREVENTION

In the same vein, might it not also be smart to reward families who engage in child-rearing practices that we know are supportive of emotional health? The child abuse data are hard for us as a nation to rationalize. Even harder to face is the fact that infants and toddlers across our nation are daily traumatized by ignorance of their need for an intimate, predictable, committed and genuinely attuned relationship. For any who doubt the degree to which this understanding is being missed, visit a child care center and watch for the signs of attachment between a little one and her child care worker. Or buy a book on "nannying," and note how many pages are devoted to the quality of the emotional bond between the nanny and the child.

Matthew is a poster child for this often overlooked form of trauma. He was four years old when his parents brought him to therapy after being kicked out of his third preschool for biting, hitting and kicking other kids. Matthew's parents made a six-figure income, but both had been traveling for work ever since Matthew was born. When they were home, they hated to spend all of their time disciplining Matthew, though they had witnessed his rageful behavior for some time.

During the first interview with the family, Matthew, in addition to literally dismantling the therapist's office, slugged his mother in the stomach, pulled her hair, and kicked and bit her. As the therapist inquired about his history, she learned that during his four short years Matthew had been in the care of fourteen alternative caregivers, five of whom had been highly paid to live in the family home. The backstory was not due, as one might assume, to Matthew's difficult behavior. This was a story of ignorance and grief—for Matthew and for his mother. On returning home from a trip, Matthew's mother would find that he seemed more attached to the nanny than to her and, heartbroken, would accuse the nanny of undermining her role as Matthew's mother and fire the nanny. This scenario played out five times in his first twenty-eight months of life. Matthew's subsequent caregivers were day care teachers in a series of high-end day care centers where he spent extended hours. Matthew soon generated his own losses of potential attachment with these latest caregivers, pushing against all efforts at affection and discipline.

This is a story of trauma, one too typically diagnosed as ADHD or Asperger's or a host of other possibilities. Certainly, Matthew was also depressed, but at the core was his continual loss of any adult with whom he had developed a bond—trauma for any infant or toddler. Therapy for this family involved intensive family work as well as individual sensorimotor work with Matthew, whose need for additional therapy continues. He is now in sixth grade and in a special behavioral classroom. As a teenager, Matthew is a likely candidate for risk-taking behavior, including substance abuse. Health outcomes will probably not be clear until Matthew is middle-aged, when the accumulation of his experiences interacting with his genetics will reveal their toll. Matthew's early story is extreme, but only in the number of losses he endured. Many well-meaning parents are ignorant about what constitutes trauma for a young child and unwittingly submit young children to situations that overwhelm their systems. This ignorance crosses all incomes, races and ethnicities.

Child care is more than day care. If we are to curb the current wave of health risks threatening our children, life needs to look very different for parents. Building healthy children from the beginning costs a fraction of what it costs to repair broken ones. Our nation's current approach to child-rearing is a variation on the old American adage, "If it ain't broke, don't fix it"—the same mentality that led to the 2010 Gulf oil spill. What might have once cost a few million dollars to build correctly took billions to fix, and the repercussions continue to play out. Drilling a second well as insurance in the beginning, carefully checking blowout preventers, providing adequate crews to do the job well, and doing safety checks along the way could have prevented the disaster.

For American families, a national and universally available "Parenting Institute" would be an opportunity to apply this understanding. (The closest existing example of this approach is Australia's Positive Parenting Program, known as "Triple P"; see appendix E for details.) A Parenting Institute would bring together key units of emotional developmental information from world-renowned experts, combined with the fundamentals of brain science. It would recruit experienced family practitioners who embody a sense of warmth, humor and joy in connecting with families. Like our model for children's physical health (well-child checks, immunizations and so on), the model would be oriented toward prevention.

Some of us remember the dawning of the Lamaze movement of half a century ago. At that time it was common practice for mothers to go into labor with little knowledge or preparation for the process. Fathers were typically excluded. Lamaze and childbirth education created an entirely new expectation: that parents would be informed and educated about the process of labor and birth. Thanks to Lamaze, the professional's role has shifted from total control of the process to a team effort that begins with shared knowledge and coaching so that parents make their own healthy

244

SCARED SICK

decisions. This is exactly the expectation that a Parenting Institute would generate for parents with children all along the developmental spectrum.

Imagine what we might achieve if we applied a pediatric health model to children's emotional health. Suppose you are the mother of a hard-to-comfort newborn or an oppositional four-year-old who fights and struggles every step of the way as you walk through the mall. Or perhaps you're the father of a thirteen-year-old girl who seemed to be doing fine until a month ago, when she suddenly began refusing to talk to you and won't come out of her room, saying she's fat and ugly and has no friends and doesn't want to go to school. Imagine that, as easily as you could schedule a haircut, you could stop by an inviting place in the mall or somewhere convenient on your beaten path to receive fifty minutes worth of help with your concern about your child, whatever it is. You could continue your visits until the situation was resolved or you were referred to a better resource.

Imagine that this place is warm and attractive, a place where everyone goes. Looking more like a Starbucks than a clinic, it is clearly not a social service agency. Everything about it is upbeat and welcoming. Here the guidance you receive is based on the best of what we know about healthy emotional development. You can meet privately with an experienced, wise person who listens and tailors a response to your particular child and family. If you attend a series of sessions, you can even earn educational credit, high school or college. If you invest in a series of well-child visits to learn how to handle upcoming changes in your child's development, you can earn a tax credit for your role in strengthening the fabric of your community. The message conveyed is simple: *all* of us who have children face challenges as children develop. It's normal. It goes with the territory. What makes a difference is applying the best of what we know at the earliest possible time.

We are seeing a precedent for this type of family service. Across the country, large drugstore chains—such as Walgreen's, CVS and Wal-Mart—have begun offering routine medical care at walk-in health clinics inside the stores. Piloted and bankrolled by business visionaries like Stephan Case, former chairman of AOL, and Michael Howe, former president of Arby's, these retail medical clinics—staffed primarily by nurse-practitioners who are linked to physicians for referrals—are being put in place as an alternative to the expense and inconvenience of visits to a doctor's office or emergency room.

As a society, we have a choice: we can continue to focus only on expanding systems that provide late-stage treatment or containment for people who suffer from trauma and related pathologies—a path becoming increasingly unaffordable—or we can begin to invest in building physically and emotionally healthy children from the beginning.

THE FOURTH TRIMESTER

Uniquely unfinished at birth, human brains are designed to need and respond to and grow inside of relationships. Like baby kangaroos, baby humans need a period of extended coddling between birth and exposure to the world. Ignorance of this need confounds many parents who stumble through the first three or four months of their new baby's life as if lost in Oz. The normalcy of the need for intensive attention to acclimate a new nervous system and the investment of time essential to that process is a message we need to promulgate in our culture. So here's a final "imagining." Imagine that rather than simply punishing parents who fail to adequately nurture their children, our nation began rewarding parents who invest in pro-social practices known to build educational and social success and health. Imagine that either a mom or dad who wants to stay home with their baby and participate in parent education is guaranteed paid parental leave at least through the child's first year. The majority of countries in the world today have laws that guarantee paid parental leave ranging from two months to four years. We are one of only four countries that don't have such a policy. We might begin funding such a program by offsetting a small percentage of the budget for remedial services, like prisons, so that we could gradually install upstream investments. We already have the proven programs and the technology. We could do everything suggested here, including a national universal parenting program, for a fraction of the price we are currently paying for attempted remediation of the same issues—not only in financial but also in human terms.

WALKING THE WALK

These are just a few of many implications of the research presented in *Scared Sick*. The good news is that the opportunities are as ubiquitous as the problems of emotional trauma; they are all around us. If we could take only one message from the research it would be the critical importance of the impact of first relationships on our health. Everything that we do with a baby—every gesture, every touch, every tone of voice—is incrementally building a tiny nervous system that will affect that child's health and well-being for a lifetime. Everything that we do to give substance to the understanding that early experience matters—in our own families, in our entertainment and media, in our systems of support for families, in our businesses, policies, insurance practices, courts, medical practices and laws—even in our daily interactions with parents in our neighborhoods—will mitigate against the health problems soaring in our country. A mother who chooses to breastfeed her infant for a month or two longer than might be easy, a grandmother who takes over so the mother can take a few hours away, an employer who allows flexible scheduling for employees who want to job-share to juggle children's needs, a store that installs baby-friendly restrooms with a comfortable chair for nursing, a legislator who becomes a champion for home visitation—there are as many ways as there are individuals to walk the walk.

OTHER PEOPLE'S KIDS

Beyond America, what is the relevance of this information to the rest of the world? Let's take a look at UNICEF data:

- Between 500 million and 1.5 billion children worldwide have been affected by violence.[5]
- Approximately 150 million children ages five to fourteen are engaged in child labor.[6]
- There are 145 million orphans in the world.[7]
- An estimated 560,000 to 700,000 children each year are involved in forced commercial sexual exploitation.[8]
- More than 250,000 children are currently serving as child soldiers.[9]
- At least 6 million children have been permanently disabled or seriously injured by armed conflict.[10]
- The prevalence of mental disorders is much higher in children who have experienced war, with approximately 50 percent exhibiting PTSD.[11]

The data provide only one lens. Stories emerging from the war zones provide a closer look. Before her murder in 2008 in Mosul, Sahar al-Haideri was a journalist with the Institute for War and Peace Reporting. She opened a 2007 feature article by describing Iraqi children five to seven years old, playing on the street: the little girls are screaming in mock terror as the little boys, brandishing wooden swords, pretend to kidnap the girls and threaten to behead them. The violence these children have witnessed is the stuff of their daily play as well as their night terrors. According to the Ministry of Education, in the four months preceding al-Haideri's observation of the children, there were 417 attacks on schools. UNICEF reports that school attendance is steadily dropping in Iraq following the deaths of 311 teachers and 64 children, and that 47 children were kidnapped on their way to school during the same four-month period. One teacher put it simply: "Our children have lost their childhood." Almost all of the children who have survived the terror have had a friend or relative killed or kidnapped. Al–Haideri said that many refer to this generation of Iraqi children as "the lost generation." When the boys talk about cars, they are discussing the models used in car bombings in their neighborhoods. Growing up without a stable base and a moral center, adolescent boys are easy prey for criminals. One boy is called "the prince" on his street. At age eighteen, he makes a living helping gang members steal cars on the streets of Baghdad. When asked why he has chosen a life of crime, he responds, "Big fish eat smaller ones. . . . We all have to live."[12]

Six-year-old Ayat Salah is the poster child for childhood trauma. She stopped talking after she found a headless body in front of her house in Baghdad. Ayat's mother said: "Ayat had kissed me and her father goodbye in the morning as usual . . . then she left the house, and suddenly we heard her scream and saw she had fainted. She hasn't spoken a word since." Given that children under eighteen are nearly half of Iraq's population, and given trauma's impact on the human brain, the stuff of their child's play today can only continue to shape their health—and our world—in the future.[13]

So why should we care beyond our own borders and our own health? Why is this our problem? *Scared Sick* reveals that people who are traumatized are going to function at their most primitive level of self-interest to secure their own survival. Maslow's hierarchy of needs applies. When we are frightened, we seek safety at all costs. Reason, ethics and values are sacrificed to meet basic needs. An individual who has experienced chronic fear or trauma early in life—or even later in life—so that his or her nervous system has been stripped to a chronic state of hypervigilance is not only more susceptible to illness but less able to employ higher rational skills in making complex decisions. Such an individual becomes vulnerable to messages of tyranny that are cloaked in messages of protection. The use of fear to gain compliance is a known tactic in warfare, brainwashing and totalitarian governments.

When we read of the impact of 9/11, a single terrifying experience, on babies in utero and then learn of the enormous scale of trauma to children from abuse in our own country and from war-driven trauma around the world, we have a snapshot of the long-term consequences to us all when millions of children are being programmed mentally and physically by early fear.

SMALL WORLD

Emotional trauma to an individual child creates a ripple effect, dysregulating protective systems, beginning in the child's body and resonating through the family. When such experiences are widespread, the effects undulate through communities and societies. Societal patterns are built from the collective moment-by-moment interactions between children and those who care for them, superseding language as one right brain connects to another. Thus, "the child is father of the man."

There is little question that fear and emotional trauma are at the root of many disease processes affecting humanity. But fear may present an even greater threat to us than physical disease. It may be that the more immediate threat to humanity is the undermining of cortical responses—rational and mindful thought, the ability to be present and sensitive, to empathize, to control strong negative emotions, to be guided by internalized values in the face of complex challenges.

"Fear makes you stupid," says leadership trainer Brian Regnier. When our minds and bodies are in a state of either acute or chronic red alert, our capacities for complex thinking and caring about and connecting to others are undermined. Both individually and collectively, we run toward whatever we perceive as the strongest and most expedient protection. History is filled with examples of fear used to control human behavior. As we confront profound differences in culture, race and religion and increasing competition for dwindling resources, balanced, nonimpulsive, thoughtful decisions have never been more crucial to human society.

We are slowly awakening to the understanding that our world is increasingly at peril from the effects of global warming. We understand that there is a connection between greenhouse gases and icebergs melting into the seas. What we don't seem to understand is that similar processes are at work inside our own bodies. Widespread trauma to developing nervous systems is potentially as catastrophic to human society as the greenhouse gases are to the planet.

The world we have created for ourselves is very different from the one for which we originally evolved. Humanity's challenges are no longer primarily about managing the planet's natural cycles. We face a greater challenge: to manage a world we ourselves have engineered, including the capacity to eradicate our own species through nuclear or biological weapons. But even as the need for higher cortical capacities increases in the twenty-first century, children across the globe are being affected by intolerable rates of trauma, which undermine these very capacities.[14] If the UNICEF data are accurate, and if we recognize that the effects of trauma can accrue over generations, what is the cumulative price to our species?

Humane societies don't emerge from a developmental void. We literally hold them on our laps and rock them in our arms. It is within protected and loving relationships that humans learn the brain-based underpinnings for empathy, for the regulation of strong negative emotions, and for complex problem-solving—capacities fundamental to addressing literally all of the other issues we care about, whether we're talking about war, hunger, racism, disease or the depletion of natural resources. The real question is not whether but where to start. Where in the midst of worldwide turbulence, financial recession, nuclear proliferation, health crises and competition for diminishing natural resources do we begin? The answer: in the beginning.

Time is short. Our world is rapidly changing. Thoughtful dialogue about our society's values, beliefs and child-rearing practices must take place now. The choices we make will have profound impact on the trajectory of our society—and our species. If we choose well, untapped potentials will emerge. If we remain passive and let the momentum of our dissolving social structures sweep us into the next generation, we lose the creativity and productivity of millions of children. And we lose our future.

—DR. BRUCE PERRY,
Biological Relativity: Time and the Developing Child

APPENDIX A

Preventing Trauma to
Young Children During Divorce

There are several steps parents can take to prevent trauma to the youngest children when a marriage is coming apart:

- Make every effort to repair the relationship. The stone you leave unturned will be the one that will haunt you and that you will have the greatest difficulty explaining to your child later. Therapy can make huge differences.
- If you believe that divorce is your only option, see a therapist before you file. A good therapist can save you lots of tears and money, is much less expensive than a lawyer, and will help you support your children's health. Create a parenting contract and keep it updated.
- Regardless of the decision about not staying together for now, remember that as parents you will be standing together at future birthdays, athletic events, graduations, weddings and births. Create a conciliatory and amiable legacy for the future. Take the high road.
- Minimize conflict and anger, especially in the child's presence but also in legal and custodial transactions. Being generous with each other is much less expensive in the long run.
- Maintain boundaries. Don't confide your problems to your children or talk negatively about the other parent or make faces indicative of your feelings (including telephone, e-mail and text messages).

- Don't fight where the kids can hear you. If your best efforts fail and the kids do overhear a fight, talk to them about it. Don't pretend nothing happened. Model what you hope your child will do in similar circumstances: take a time-out when you can't handle your anger and return only when you are stable.
- Do everything possible to protect a warm, secure, stable attached relationship with your child and encourage the same with the other parent.
- Maintain or create predictable routines that include both parents whenever possible. This doesn't mean every day is equally divided between parents but rather that over time the child has flexible time with both parents, as developmentally appropriate.

APPENDIX B

Maltreatment in Childhood

Having been mistreated as a child correlates with the following issues in adults[1]:

Alcohol abuse and illicit drug use
Obesity (particularly correlated with childhood sex abuse)
Risky sexual behaviors
Cigarette smoking
PTSD and dissociative disorders
Depression: self-harm and suicide
Anxiety
Panic disorder
Genitourinary disorders
Skin problems
Respiratory illness
Muscular-skeletal disorders (for instance, headaches)
Memory loss/Alzheimer's
Chronic pain
Ischemic heart disease
Liver disease
Digestive disorders
Asthma and allergies

APPENDIX C

Recognizing Trauma

EARLY SIGNS OF EMOTIONAL TRAUMA
IN PRESCHOOLERS

Neglect

A neglected child may show the following behaviors:

- Developmental delays (motor, language, social, cognitive)
- Odd eating behaviors (hoarding, hiding, failure to thrive, swallowing problems, gluttony, rumination, throwing up)
- Primitive self-soothing or stimulating behaviors (rocking, headbanging, scratching, cutting, chanting or other repetitive sounds that increase when stressed)
- Indiscriminate attachment (affection with anyone)

Abuse

In addition to the behaviors that can indicate neglect, an abused child may also:

- Replicate abusive behaviors, whether physical (for example, hitting), verbal (cruelty and inappropriate language), or sexual (sexual behavior or sexualized language beyond the child's maturity)
- Be aggressive
- Have sleep disturbances (nightmares and night terrors)

- Exhibit regressive behavior (bed-wetting, clinginess, loss of speech, fear of new situations)

Depression

A child who is depressed may experience or manifest the following:

- Bouts of sadness (interspersed with periods of normal behavior)
- Lack of pleasure in play
- Sleep disturbance
- Acting out/aggression

SYMPTOMS OF EMOTIONAL TRAUMA IN LATER DEVELOPMENT

In grade school children and teenagers, emotional trauma may manifest in the following ways:

Physical Symptoms

- Eating disturbances (more or less than usual); excessive weight gain
- Sleep disturbances (more or less than usual); nightmares or night terrors
- Excessive or undiscriminating, provocative sexual behavior (masturbation, self-stimulation, preoccupation with sex or seduction)
- Low energy
- Chronic, unexplained pain

Emotional Symptoms

- Depression, spontaneous crying, despair and hopelessness
- Anxiety
- Panic attacks
- Fearfulness
- Compulsive and obsessive behaviors
- Feeling out of control

- Irritability, anger and resentment; unpredictable aggressive outbursts
- Emotional numbness
- Withdrawal from normal routines and relationships

Cognitive Symptoms

- Memory lapses, especially about trauma
- Difficulty making decisions
- Decreased ability to concentrate
- Distractibility; ADHD symptoms; extreme "dreaminess" (dissociation)

APPENDIX D

Working with Traumatized Children

Traumatized children are children who have adapted by adjusting their emotions, thinking and basic physiological responses to prepare for the worst. They are less able to concentrate and are hyper-alert to nonverbal cues like tone of voice, gestures and facial expressions. They are employing either hyper-arousal or dissociation to survive.

GUIDELINES FOR CAREGIVERS

1. Provide structure and a predictable and consistent rhythm for the day: Let the child know about new or different activities beforehand whenever possible. A printed calendar recording the daily and weekly plans may be helpful. The more time a child has to adjust to any change, the better. Any perception that the adults in charge are in any way disorganized or confused or anxious will add to the insecurity of traumatized children.

2. Don't be reluctant to talk about the trauma: When children sense that caregivers are upset, it only makes things worse. Listen, answer questions and offer support and comfort. Don't avoid the topic of the trauma, but don't overreact to it either. Let the child lead. Avoidance or pretending about the circumstances only makes the trauma worse in a child's eyes.

3. Be nurturing, affectionate and comforting to the degree that is comfortable for the child: Intimacy can be a source of confusion for children traumatized by physical or sexual abuse. Hugs and kisses for youngest children are important. But take your cues from the child. Provide affection when a child seeks it or is receptive. If he or she wants to be held or rocked, do so, but don't impose affection on a reluctant child.

4. Discuss your expectations and discipline plan with a child: Clear rules and the consequences of breaking them need to be understood in advance. Whatever system you use needs to be discussed ahead of time— maybe more than once—so that the child feels that he or she can predict what will happen and feels things are "fair." The child will test you, so consistency is critical. Your system needs to be built on positive reinforcement and rewards, not punishment. Avoid physical discipline and learn constructive time-out—an opportunity for a child to stop, calm and regulate for a few minutes in a safe setting, not be isolated or shunned.

5. Give the child choices and a sense of control: For a child given some degree of control and constructive choices to make, feelings of safety and comfort will be enhanced and acting-out will be reduced. Frame the consequence of noncompliant behavior as a choice: "You can follow the rule or you can choose something else which you know will lead to. . . . "

6. Talk with the child whenever you can: Give him or her age-appropriate information about how the adult world works. Traumatized children's fears and fantasies are much more frightening than the truth. Honesty and openness are key to trust. If you don't know the answer to a child's question, it's okay to say, "I don't know."

7. Watch for signs of re-enactment: In play, in artwork, in behaviors in general, you may notice signs of dissociation (avoidance of other children, daydreaming, withdrawal) or of hypervigilance (anxiety, impulsivity or sleep problems.) These behaviors are typical for traumatized children and indicate that they are being reminded of the event, either through internal processes (thoughts or flashbacks) or experiences that re-stimulate the memory. Comfort and tolerance are the best approach to re-enactment, unless the child victimizes someone else. These symptoms are likely to ebb and flow, often with no apparent provocation. Try to observe patterns and cues that may help therapeutically.

8. Protect the child: Don't hesitate to stop activities that upset the child. Avoid or limit movies, television shows, games or any other activity that re-stimulates a child's symptoms.

9. Don't be afraid to ask for help! Knowledge is power. A trained professional may be your right arm in re-setting the course for a troubled child.

Adapted from Bruce D. Perry, "Principles of Working with Traumatized Children," available at: http://teacher.scholastic.com/professional/bruceperry/working_children.htm.

APPENDIX E

Preventive Programs Proven Effective

There are many programs that have shown promise in preventing trauma to young children in the United States. The following are those that have passed the "gold standard" of rigorous long-term evaluation, resulting in replicable protocols for natural application.

LONG-TERM OUTCOMES FROM THE NURSE-FAMILY PARTNERSHIP HOME VISITATION PROGRAM

The Nurse-Family Partnership model began in Elmira, New York, in 1976 with the first of three randomized, controlled trials. Two additional research sites were added later, in Memphis, Tennessee (1988), and Denver, Colorado (1994).

The NFP is an evidence-based community health program that helps to transform the lives of vulnerable mothers who are pregnant with their first child. Each mother served is partnered with a registered nurse early in her pregnancy, and she receives ongoing visits that continue through her child's second birthday. Follow-up research on the long-term outcomes for mothers and children in these three trials continues to this day. NFP is the most thoroughly evaluated home visitation program in the nation and is considered to be the gold standard.

The following outcomes have been documented among participants in at least one of the trial sites:

IMPROVED PREGNANCY OUTCOMES

- A 79 percent reduction in preterm delivery for women who smoke
- Reductions in high-risk pregnancies as a result of greater intervals between first and subsequent births

IMPROVED CHILD HEALTH AND DEVELOPMENT

- 59 percent reduction in child arrests at age fifteen
- 39 percent fewer injuries among children
- 56 percent reduction in emergency room visits for accidents and poisonings
- 48 percent reduction in child abuse and neglect
- 50 percent reduction in language delays of children at twenty-one months
- 67 percent reduction in behavioral and intellectual problems at age six

INCREASED ECONOMIC SELF-SUFFICIENCY

- 32 percent fewer unintended subsequent pregnancies
- 83 percent increase in mother's labor force participation by the child's fourth birthday
- 20 percent reduction in months on welfare
- 46 percent increase in father's presence in household
- 60 percent fewer arrests of the mother
- 72 percent fewer convictions of the mother

Adapted from "Nurse-Family Partnership Fact Sheet: Research Trials and Outcomes," 2010, available at: http://www.nursefamilypartnership.org/assets/PDF/Fact-sheets /NFP_Research_Outcomes.

Readers interested in more information can contact the Nurse-Family Partnership National Service Office at:
  1900 Grant Street, Suite 400
  Denver, CO 80203
  303-327-4240; (toll-free) 866-864-5226
  www.nursefamilypartnership.org

NURSE FAMILY PARTNERSHIP:
HOSPITALIZATIONS FOR WHICH INJURIES OR INGESTIONS WERE DETECTED
(FOR MEMPHIS NFP RESEARCH SITE)

CHILDREN WHOSE MOTHERS WERE VISITED BY NURSES PRENATALLY UP TO AGE 2		CHILDREN WHOSE MOTHERS WERE NOT VISITED BY NURSES	
DIAGNOSIS	DAYS IN HOSPITAL	DIAGNOSIS	DAYS IN HOSPITAL
Burns	2	Head trauma	1
Coin Ingestion	1	Fractured fibula/congenital syphilis	12
Ingestion of iron medication	4	Strangulated hernia with delay in seeking care/burns	15
		Bilateral subdural hematoma	19
		Fractured skull	5
		Bilateral subdural hematoma/aseptic meningitis, 2nd hospitalization	4
		Fractured skull	3
		Coin ingestion	2
		Child abuse/neglect suspected	2
		Fractured tibia	2
		Burns (2nd and 3rd degree to face/neck)	5
		Burns (2nd and 3rd degree to bilateral leg)	4
		Gastroenteritis/head trauma	3
		Burns (splinting/grafting), 2nd hospitalization	6
		Finger injury/osteomyelitis	6
TOTAL DAYS HOSPITALIZED	7	TOTAL DAYS HOSPITALIZED	89

The chart above is a sample compilation of such research data, a record of hospitalizations for injuries or ingestions detected among children at the Memphis site:

HEALTHY FAMILIES AMERICA HOME VISITATION

Healthy Families America (HFA) benefits children and families through positive child health and development, enhanced parenting and reduced child abuse and neglect. The research support for HFA comes from thirty-four evaluations in twenty-five states, with varying results. HFA sites serve many types of families using a variety of curricula, and that variation has an impact on the strength of evaluation results. The results from Healthy Families New York and Oregon's Healthy Start are particularly promising:

Birth Outcomes and Birth Weight: Two rigorous studies show improvement in birth outcomes of 55 percent or more, including more babies born at healthy weights and fewer babies with birth complications.

Breast-feeding: HFA boosted rates of breast-feeding by 25 percent or more in two studies, for parents enrolled prenatally. Breast-feeding has demonstrated significant positive impacts on child and maternal health.

Parenting Attitudes: Parenting attitudes improved faster for those enrolled in HFA compared to those in a control group in the majority of studies. The research literature shows a clear relationship between parenting attitudes and child maltreatment.

Attachment and Home Environment: The vast majority of studies at all levels of rigor show significant improvements in home environment, including parent-child interaction and developmental stimulation.

Interpersonal Violence: HFA mothers showed significant decreases in perpetration of and victimization from physical assault—34 percent and 21 percent, respectively.

Child Abuse and Neglect: A large study found less physical and psychological abuse for HFA parents than for control parents at one year. Results at two years showed the greatest impacts for first-time mothers and psychologically vulnerable mothers.

CONTACTS

Healthy Families America
228 S. Wabash, 10th Floor
Chicago, IL 60604
312-663-3520
www.healthyfamiliesamerica.org

Healthy Families New York
Offices of Children and Families
52 Washington Street, 3rd Floor
North Rensselaer, NY 12144
518-474-9486
www.healthyfamiliesnewyork.org

Healthy Start of Oregon
Christi Peeples, Development and Prevention Services Coordinator
503-378-6768
Christi.Peeples@state.or.us

Research summary provided by Kathryn Harding, Prevent Child Abuse America (June 2010).

TRIPLE P
(POSITIVE PARENTING PROGRAM)

The Positive Parenting Program, or "Triple P," is a system for delivering parenting tools and techniques to parents for constructive child behavior management. It offers a range of options from newsletter articles to brief consultations to ten weeks of parent coaching. Founded in Queensland, Australia, thirty years ago, Triple P is grounded in extensive research and has now spread to numerous countries around the world. In the United States, beginning in 2003, it has been tested in several sites in South Carolina.

The program is a universal model and designed to be accessible to all parents within a given community. Rather than singling out "high-risk parents" for interventions, all parents are provided with multiple levels of parenting support of increasing intensity to match each family's needs. The South Carolina study included eighteen counties, nine of which were chosen randomly for the parenting intervention, with a total of approximately 100,000 children ages birth to eight years in the nine Triple P counties. Counties that provided Triple P services (by contrast to those nine that did not) saw:

- A 25 percent reduction in abuse and neglect
- A 33 percent reduction in foster care placements
- A 35 percent reduction in emergency room visits or hospitalizations for abuse or neglect

There are five core Triple P principles:

1. Ensuring a safe, engaging environment
2. Promoting a positive learning environment
3. Using assertive discipline
4. Maintaining reasonable expectations
5. Taking care of oneself as a parent

ADDITIONAL EARLY CHILDHOOD HOME-VISITING RESOURCES

Several other well-established home-visiting programs are designed to maximize early development.

Early Head Start is a federally funded community-based program for low-income pregnant women and families with infants or toddlers designed to promote healthy prenatal outcomes and enhance the development of very young children.

Early Head Start National Resource Center
2000 M Street, NW, Suite 200
Washington, DC 20036
202-638-1144
www.ehsnrc.org

The **Parent Child Home Program** is an early childhood literacy, parenting and school-readiness program. It uses trained paraprofessionals to work with families who have not had access to educational and economic opportunities. It prepares children for academic success and strengthens families through intensive home visiting.

Parent Child Home Program
1415 Kellum Place, Suite 101

Garden City, NY 11530
516-883-7480
www.parent-child.org

The **Home Instruction Program for Preschool Youngsters** (HIPPY) is a home-based, family-focused program that helps parents of three-, four- and five-year-olds provide educational enrichment for their child to prepare them for school.
HIPPY USA
1221 Bishop Street
Little Rock, AR 72202
501-537-7726
www.hippyusa.org

Parents as Teachers (PAT) is a parent education and family support program serving families, beginning with pregnancy until the child enters kindergarten. PAT-certified parent-educators are trained to translate scientific information on early brain development into specific *when, what, how* and *why* advice for families.
Parents as Teachers National Center
2228 Ball Drive
St. Louis, MO 63146
314-432-4330
www.parentsasteachers.org

NOTES

Introduction

1. J. Warner, "Dysregulation Nation," *New York Times Magazine*, June 20, 2010.

2. "U.S. Life Expectancy Lags Behind 41 Nations," *USA Today*, August 11, 2007.

3. C. Fryar, R. Hirsch, M. Eberhardt, S. Sug Yoon, and J. Wright, "Hypertension, High Serum Total Cholesterol, and Diabetes: Racial and Ethnic Prevalence Differences in U.S. Adults, 1999–2006," NCHS data brief, April 2010, available at: http://www.ncbi.nlm.nih .gov/pubmed/20423605 (accessed July 15, 2011).

4. L. E. Fields, V. L. Burt, J. A. Cutler, J. Hughes, E. J. Roccella, and P. Sorlie, "The Burden of Adult Hypertension in the United States, 1999 to 2000: A Rising Tide," *Hypertension* 44, no. 4 (October 2004): 398–404.

5. Ibid.

6. P. Belluck, "Obesity Rates Hit Plateau in U.S., Data Suggest," *New York Times*, January 13, 2010.

7. Centers for Disease Control and Prevention, Office of the Association Director for Communication, "Number of People with Diabetes Increases to 24 Million," press release, June 24, 2008, available at: http://www.cdc.gov/media/pressrel/2008/r080624.htm (accessed April 28, 2010).

8. National Institute of Mental Health (NIMH), "The Numbers Count: Mental Disorders in America," 2010, available at: http://www.nimh.nih.gov/health/publications/the-numbers-count-mental-disorders-in-america/index.shtml#ADHD (accessed June 23, 2010).

9. Ibid.

10. B. F. Grant, F. S. Stinson, D. A. Dawson, et al., "Prevalence and Co-occurrence of Substance Use Disorders and Independent Mood and Anxiety Disorders: Results from the National Epidemiologic Survey on Alcohol and Related Conditions," *Archives of General Psychiatry* 61, no. 8 (August 2004): 807–816.

11. *USA Today*/HBO, drug addiction poll, May 2006, available at: http://www.hbo .com/addiction/understanding_addiction/17_usa_today_poll.html (accessed July 15, 2011).

12. Every Child Matters Education Fund, "Homeland Insecurity," Washington, D.C., April 2009.

13. Ibid.

14. Centers for Disease Control and Prevention, Office of the Association Director for Communication, "Number of People with Diabetes Increases to 24 Million," June 24, 2008.

15. Every Child Matters Education Fund, "Homeland Insecurity."

16. U.S. Department of Health and Human Services (DHHS), Administration on Children, Youth, and Families (ACYF), "Child Maltreatment," Washington, D.C., 2007.

17. S. Hensley, "Kids Become Prime Growth Market for Prescription Drugs," National Public Radio, May 19, 2010, available at: http://www.npr.org/blogs/health/2010/05/19/126975784/kids-become-prime-market-for-prescription-drugs?sc=17&f=1001.

18. Pew Center for the States, "Delivering Healthier Babies and Economic Returns," Washington, D.C., November 2009.

19. National Institute of Mental Health (NIMH), "Any Disorder Among Children," 2010, available at: http://www.nimh.nih.gov/statistics/1ANYDIS_CHILD.shtml (accessed July 15, 2011).

20. SAMHSA, "Helping Children and Adolescents Who Have Experienced Traumatic Events," May 3, 2011, available at: http://digitallibraries.macrointernational.com/gsdl/collect/cmhsdigi/index/assoc/HASH01f9.dir/doc.pdf (accessed July 12, 2011).

21. T. Maugh, "The Heart Disease Trifecta," *Los Angeles Times*, April 26, 2010.

22. Centers for Disease Control and Prevention, *Summary Health Statistics for U.S. Children: National Health Interview Survey, 2006*, Washington, D.C.: U.S. Department of Health and Human Services, September 2007.

Chapter 1

1. "Oprah Talks Weight Gain," January 5, 2009, available at: http://www.celebrity-gossip.net/celebrities/hollywood/oprah-talks-weight-gain-210278 (accessed July 15, 2011).

2. O. Winfrey, "How Did I Let This Happen Again?" *O: The Oprah Magazine*, January 2009.

3. M. Rosen, "Big Gain, No Pain," *People*, January 14, 1991.

4. P. Belluck, "Obesity Rates Hit Plateau in U.S., Data Suggests," *New York Times*, January 13, 2010.

5. S. Earls, "In the Sweet Land of Liberty . . . One Nation, Overfed: An Indulgent Culture Faces a Self-Inflicted Health Crisis," *Times Union*, January 31, 2006.

6. Centers for Disease Control and Prevention, Office of the Association Director for Communication, "Number of People with Diabetes Increases to 24 Million," press release, June 24, 2008, available at: http://www.cdc.gov/media/pressrel/2008/r080624.htm (accessed April 28, 2010).

7. T. Maugh, "The Heart Disease Trifecta," *Los Angeles Times*, April 26, 2010.

8. Centers for Disease for Control and Prevention, "FastStats—Arthritis," 2010, available at: http://www.cdc.gov/nchs/fastats/arthrits.htm (accessed April 28, 2010).

9. P. Parker, "Obese Teens Turn to Surgery as Last Resort," *Oregonian*, February 24, 2008.

10. Ibid.

11. A. Finley, "Suffer the Little Children," *The Globe and Mail*, December 28, 2005.

12. Ibid.

13. J. Spencer, "I Drank the Way I Ate," *Wall Street Journal*, July 18, 2006.

14. Barbara Thompson's Weight Loss Surgery Center, available at: http://www.wlscenter.com (accessed July 14, 2009).

15. J. Spencer, "Alcoholism in People Who Had Weight-Loss Surgery Offers Clues to Roots of Dependency," *Wall Street Journal*, July 18, 2006.

16. Ibid.

17. J. Ruttiann, "Viewing Obesity as an Addiction," *Endocrine News: A Publication of the Endocrine Society* 33, no. 5 (May 2008): 23.

18. A. Goodman, "In the Realm of Hungry Ghosts: Interview with Dr. Gabor Maté," *Democracy Now!* February 3, 2010, available at: http://www.democracynow.org/2010/2/3/addiction (accessed April 28, 2010).

19. Dr. Vincent Felitti, interview with Robin Karr-Morse, San Francisco, March 7, 2005.

20. V. Felitti, "The Relation Between Adverse Childhood Experiences and Adult Health: Turning Gold into Lead," *Permanente Journal* 6 (2002): 44–47.

21. Dr. Vincent Felitti, phone interview with Robin Karr-Morse, May 10, 2007.

22. T. Bentley, and C. S. Widom, "A 30-Year Follow-up of the Effects of Child Abuse and Neglect on Obesity in Adulthood," *Obesity* 17, no. 10 (October 2009): 1900–1905.

23. Ibid.

24. D. W. Brown, R. F. Anda, H. Tiemeier, et al., "Adverse Childhood Experiences and the Risk of Premature Mortality," *American Journal of Preventive Medicine* 37, no. 5 (November 2009): 389–396.

25. J. Stevens, "Traumatic Childhood Takes 20 Years Off Life Expectancy," October 6, 2009, KTKA.com, available at: http://www.ktka.com/news/2009/oct/06/traumatic_childhood_takes_20_years_life_expectancy/.

26. Felitti, "The Relation Between Adverse Childhood Experiences and Adult Health."

27. S. R. Dube, V. J. Felitti, M. Dong, D. P. Chapman, W. H. Giles, and R. F. Anda, "Childhood Abuse, Neglect, and Household Dysfunction and the Risk of Illicit Drug Use: The Adverse Childhood Experiences Study," *Pediatrics* 111, no. 3 (March 2003): 564–572.

28. S. Dube, R. Anda, V. Felitti, D. P. Chapman, D. F. Williamson, and W. H. Giles, "Household Dysfunction and the Risk of Attempted Suicide Throughout the Lifespan," *Journal of the American Medical Association* 286 (2001): 3089–3096.

29. V. J. Felitti, "Reverse Alchemy in Childhood: Turning Gold into Lead," *Family Violence Prevention Fund: Health Alert* 8, no. 1 (Summer 2001).

30. Dr. Vincent Felitti, interview with Robin Karr-Morse, San Francisco, March 7, 2005.

31. Dube, Felitti, et al., "Childhood Abuse, Neglect, and Household Dysfunction and the Risk of Illicit Drug Use."

32. Dr. Vincent Felitti, interview with Robin Karr-Morse, San Francisco, March 7, 2005.

33. Dr. Vincent Felitti, phone interview with Robin Karr-Morse, September 17, 2007.

34. Dr. Robert Anda, phone interview with the authors, December 13, 2007.

35. Centers for Disease Control and Prevention, "Overweight and Obesity: 2005," available at: http://www.cdc.gov/nccdphp/dnpa/obesity/.

Chapter 2

1. Dr. Bruce S. McEwen, interview with the authors, New York City, April 23, 2008.

2. B. S. McEwen and E. Norton, *The End of Stress as We Know It*, Washington, D.C.: National Academies Press, 2002.

3. Salk Institute, "Possible Mechanistic Link Between Stress and the Development of Alzheimer Tangles," *ScienceDaily*, June 15, 2007, available at: http://www.sciencedaily .com/releases/2007/06/070614155344.htm (accessed July 14, 2010).

4. A. Rogers, *The Hidden Language of Trauma*, New York: Random House, 2006.

5. Santa Barbara Graduate Institute Center for Clinical Studies and Research and Los Angeles County Early Identification and Intervention Group, "Emotional and Psychological Trauma: Causes and Effects, Symptoms and Treatment," reprinted from Helpguide.org, 2005, available at: http://www.healingresources.info/emotional_trauma _overview.htm.

6. M. J. Meaney, S. Bhatnagar, S. Larocque, et al., "Early Environment and the Development of Individual Differences in the Hypothalamic-Pituitary-Adrenal Stress Response," in *Severe Stress and Mental Disturbance in Children*, ed. C. R. Pfeffer, Washington, D.C.: American Psychiatric Press, 1996, 85–131.

7. C. Gorman, "The Science of Anxiety: Why Do We Worry Ourselves Sick? Because the Brain Is Hardwired for Fear, and Sometimes It Short-Circuits," *Time*, June 10, 2002, 46–54.

Chapter 3

1. Dr. Vincent Felitti, interview with Robin Karr-Morse, San Francisco, March 7, 2005.

2. J. LeDoux, *The Emotional Brain*, New York: Simon & Schuster, 1996.

3. B. S. McEwen and E. Norton, *The End of Stress as We Know It*, Washington, D.C.: National Academies Press, 2002.

4. E. Sternberg, *The Balance Within: The Science Connecting Health and Emotions*, New York: W. H. Freeman, 2000.

5. G. Cowley and C. Kalb, "Our Bodies, Our Fears," *Newsweek*, March 3, 2003.

6. J. D. Bremner, "Alterations in Brain Structure and Function Associated with Post-Traumatic Stress Disorder," *Seminars in Clinical Neuropsychiatry* 4, no. 4 (October 1999): 249–255.

7. M. Vythilingam, C. Heim, J. Newport, et al., "Childhood Trauma Associated with Smaller Hippocampal Volume in Women with Major Depression," *American Journal of Psychiatry* 159, no. 12 (December 2002): 2072–2080.

8. R. Scaer, *The Trauma Spectrum: Hidden Wounds and Human Resiliency*, New York: W. W. Norton, 2005.

9. Ibid.

10. Dr. Robert Scaer, phone interview with Robin Karr-Morse, February 12, 2008.

11. Scaer, *The Trauma Spectrum*, 62–64.

12. Dr. Robert Scaer, phone interview with Robin Karr-Morse, February 12, 2008.

13. B. A. van der Kolk, "The Body Keeps the Score: Memory and the Evolving Psychobiology of Posttraumatic Stress," *Harvard Review of Psychiatry* 1, no. 5 (January–February 1994): 253–265.

14. J. D. Bremner, "Does Stress Damage the Brain?" *Biological Psychiatry* 45, no. 7 (April 1, 1999): 797–805.

15. Salk Institute, "Possible Mechanistic Link Between Stress and the Development of Alzheimer Tangles," *ScienceDaily*, June 15, 2007, available at: http://www.sciencedaily .com/releases/2007/06/070614155344.htm (accessed July 14, 2010).

16. N. Klein, *The Shock Doctrine: The Rise of Disaster Capitalism*, New York: Metropolitan Books, 2007.

Chapter 4

1. Dr. Robert Scaer, phone interview with Robin Karr-Morse, February 12, 2008.

2. D. Chamberlain, "What Babies Are Teaching Us About Violence," *Pre- and Perinatal Psychology Journal* 10, no. 2(September 1995): 51–75.

3. M. A. Paul, "The First Ache," *New York Times Sunday Magazine*, February 10, 2008, 46–49.

4. Ibid.

5. Ibid.

6. Ibid.

7. "Media Spotlight: Fetal Growth and Adult Heart Disease," *Journal Watch*, available at: http://pediatrics.jwatch.org/cgi/content/full/2008/109/1 (accessed April 21, 2009).

8. Ibid.

9. "Study Finds Surprising Links Between Depression, Suicide, and Epilepsy," *ScienceDaily*, October 10, 2005, available at: http://www.sciencedaily.com/releases/2005/10/051010085822.htm (accessed July 1, 2009).

10. R. Yirmiya, I. Goshen, A. Bajayo, et al., "Depression Induces Bone Loss Through Stimulation of the Sympathetic Nervous System," *Proceedings of the National Academy of Sciences USA* 103, no. 45 (November 7, 2006): 16876–16881.

11. P. Nathanielsz, *Life in the Womb: The Origin of Health and Disease*, Ithaca, N.Y.: Promethean Press, 1999.

12. J. K. Buitelaar, A. C. Huizink, E. J. Mulder, P. G. de Medina, and G. H. Visser, "Prenatal Stress and Cognitive Development and Temperament in Infants," *Neurobiology of Aging* 24, supp. 1 (May–June 2003): S53–S60; discussion S67–S68.

13. P. Stien and J. Kendall, *Psychological Trauma and the Developing Brain: Neurologically Based Interventions for Troubled Children*, Binghamton, N.Y.: Haworth Maltreatment and Trauma Press, 2004.

14. Ibid.

15. Dr. Thomas Fleming, interview with Robin Karr-Morse, July 24, 2008.

16. N. Shanks and S. L. Lightman, "The Maternal-Neonatal Neuro-Immune Interface: Are There Long-Term Implications for Inflammatory or Stress-Related Disease?" *Journal of Clinical Investigation* 108, n. 11 (December 2001): 1567–1573.

17. Nathanielsz, *Life in the Womb*, 94.

18. R. Restak, *The Infant Mind*, New York: Doubleday, 1986.

19. M. P. Austin, L. R. Leader, and N. Reilly, "Prenatal Stress, the Hypothalamic-Pituitary-Adrenal Axis, and Fetal and Infant Neurobehavior," *Early Human Development* 81, no. 11 (November 2005): 917–926.

20. A. Finley, "Suffer the Little Children," *The Globe and Mail*, December 28, 2005.

21. P. Thomson, "'Down Will Come Baby': Prenatal Stress, Primitive Defenses, and Gestational Dysregulation," *Journal of Trauma and Dissociation* 8, no. 3 (2007): 85–113.

22. Ibid.

23. Ibid., 89.

24. Nathanielsz, *Life in the Womb*.

25. Ibid.

26. Ibid., 124.

27. "Fetus to Mom: You're Stressing Me Out!" WebMD feature, MedicineNet.com, 2005, available at: http://www.medicinenet.com/script/main/art.asp?articlekey=51730 (accessed July 1, 2009).

28. R. Yehuda, S. M. Engel, S. R. Brand, J. Seckl, S. M. Marcus, and G. S. Berkowitz, "Transgenerational Effects of Posttraumatic Stress Disorder in Babies of Mothers Exposed to the World Trade Center Attacks During Pregnancy," *Journal of Clinical Endocrinology and Metabolism* 90, no. 7 (July 2005): 4115–4118.

29. Ibid.

30. Ibid.

31. N. M. Talge, C. Neal, and V. Glover, "Antenatal Maternal Stress and Long-Term Effects on Child Neurodevelopment: How and Why?" *Journal of Child Psychology and Psychiatry* 48, nos. 3–4 (March–April 2007): 245–261.

32. "Fetus to Mom: You're Stressing Me Out!"

33. Buitelaar et al., "Prenatal Stress and Cognitive Development and Temperament in Infants."

34. K. Bergman, P. Sarkar, T. G. O'Connor, N. Modi, and V. Glover, "Maternal Stress During Pregnancy Predicts Cognitive Ability and Fearfulness in Infancy," *Journal of the American Academy of Child and Adolescent Psychiatry* 46, no. 11 (November 2007): 1454–1463.

35. Ibid.

36. E. J. Mulder, P. G. Robles de Medina, A. C. Huizink, B. R. van den Bergh, J. K. Buitelaar, and G. H. Visser, "Prenatal Maternal Stress: Effects on Pregnancy and the (Unborn) Child," *Early Human Development* 70, nos. 1–2 (December 2002): 3–14.

37. M. Rizzo, "Stress in Pregnancy Linked with Schizophrenia," *BMC Psychiatry*, August 21, 2008, available at: http://www.medicineonline.com/news/12/13124/Stress-in-pregnancy-linked-with-schizophrenia.html (accessed October 26, 2008).

38. A. S. Khashan, K. M. Abel, R. McNamee, et al., "Higher Risk of Offspring Schizophrenia Following Antenatal Maternal Exposure to Severe Adverse Life Events," *Archives of General Psychiatry* 65, no. 2 (February 2008): 146–152.

39. T. G. O'Connor, J. Heron, and V. Glover, "Antenatal Anxiety Predicts Child Behavioral/Emotional Problems Independently of Postnatal Depression," *Journal of the American Academy of Child and Adolescent Psychiatry* 41, no. 12 (December 2002): 1470–1477.

40. Mulder et al., "Prenatal Maternal Stress."

41. Talge et al., "Antenatal Maternal Stress and Long-Term Effects on Child Neurodevelopment."

42. M. D. Kogan, S. J. Blumberg, L. A. Schieve, et al., "Prevalence of Parent-Reported Diagnosis of Autism Spectrum Disorder Among Children in the U.S., 2007," *Pediatrics* 124, no. 5 (November 2009): 1395–1403.

43. T. Field, M. Diego, and M. Hernandez-Reif, "Prenatal Depression Effects on the Fetus and Newborn: A Review," *Infant Behavioral Development* 29, no. 3 (July 2006): 445–455.

44. D. W. Haley, N. S. Handmaker, and J. Lowe, "Infant Stress Reactivity and Prenatal Alcohol Exposure," *Alcoholism: Clinical and Experimental Research* 30, no. 12 (December 2006): 2055–2064.

45. Mulder et al., "Prenatal Maternal Stress."

46. J. Martin, B. Hamilton, P. Sutton, et al., *Births: Final Data for 2005*, Atlanta: Centers for Disease Control and Prevention, December 2007.

47. H. Als, "Individualized Developmental Care for Preemie Infants," *Encyclopedia on Early Childhood Development*, July 2004.

48. Chuck Green, interview with Robin Karr-Morse, February 11, 2010.

49. Board of Trustees, University of Arkansas for Medical Sciences, "Teleconferences and Guidelines for Best Practices," 2009, available at: http://www.uams.edu/cdh1/guidelines .asp (accessed June 19, 2010).

50. Paul, "The First Ache."

51. A. Mitchell and B. J. Boss, "Adverse Effects of Pain on the Nervous Systems of Newborns and Young Children: A Review of the Literature," *Journal of Neuroscience Nursing* 34, no. 5 (October 2002): 228–236.

52. "Caroline," interview with the authors, February 10, 2009.

53. S. Wust, S. Entringer, I. S. Federenko, W. Schlotz, and D. H. Hellhammer, "Birth Weight Is Associated with Salivary Cortisol Responses to Psychosocial Stress in Adult Life," *Psychoneuroendocrinology* 30, no. 6 (July 2005): 591–598.

54. Martin et al., *Births: Final Data for 2005*.

55. P. Levine and M. Kline, *Trauma Through a Child's Eyes: Awakening the Ordinary Miracle of Healing: Infancy Through Adolescence*, Berkeley, Calif.: North Atlantic Books, 2007.

56. Ibid.

57. R. Scaer, *The Trauma Spectrum: Hidden Wounds and Human Resiliency*, New York: W. W. Norton and Co., 2006.

58. Martin et al., *Births: Final Data for 2005*.

59. Childbirth Connection, "Why Does the National U.S. Cesarean Section Rate Keep Going Up?" *Cesarean Section*, 2010, available at: http://www.childbirthconnection.org /article.asp?ck=10456 (accessed June 10, 2010).

60. M. Miller, "Study: Preemies Face Risks as They Grow Up," CBS News, March 25, 2008.

61. K. J. Anand and F. M. Scalzo, "Can Adverse Neonatal Experiences Alter Brain Development and Subsequent Behavior?" *Biology of the Neonate* 77, no. 2 (February 2000): 69–82.

62. Ibid.

63. Ibid.

Chapter 5

1. "Time Bomb," *US Weekly*, January 21, 2008.

2. Ibid.

3. Wanda Kaczynski, interview with the authors, Schenectady, N.Y., November 8, 2008.

4. A. Rogers, *The Hidden Language of Trauma*, New York: Random House, 2006.

5. P. Stien and J. Kendall, *Psychological Trauma and the Developing Brain: Neurologically Based Interventions for Troubled Children*, Binghamton, N.Y.: Haworth Maltreatment and Trauma Press, 2004.

6. L. Young and T. Insel, "Hormones and Parental Behavior," in *Behavioral Endocrinology*, 2nd ed., ed. J. Becker, S. M. Breedlove, D. Crews, and B. Books, Cambridge, Mass.: MIT Press, 2002.

7. R. Avitsur, J. Hunzeker, and J. F. Sheridan, "Role of Early Stress in the Individual Differences in Host Response to Viral Infection," *Brain, Behavior, and Immunity* 20, no. 4 (July 2006): 339–348.

8. A. F. Lieberman and K. Knorr, "The Impact of Trauma: A Developmental Framework for Infancy and Early Childhood," *Pediatric Annals* 36, no. 4 (April 2007): 209–215.

9. J. Miller, "Infants in Danger: A Conversation with Infant Psychologist Alicia Lieberman," UCSF, May 2, 2008, available at: http://www.ucsf.edu/science-cafe/conversations/infants-in-danger-a-conversation-with-infant-psychologist-alicia-lieberman (accessed June 19, 2010).

10. J. Wallerstein, J. Lewis, and S. Blakeslee, *The Unexpected Legacy of Divorce: A 25-Year Landmark Study*, New York: Hyperion, 2000.

11. E. M. Hetherington and J. Kelly, *For Better or Worse: Divorce Reconsidered*, New York: W. W. Norton and Co., 2002.

12. J. Warner, "Dysregulation Nation," *New York Times*, June 20, 2010.

13. Children and Youth in History, "UNICEF Data on Orphans by Region to 2010," http://chnm.gmu.edu/cyh/primary-sources/293.

14. NumberOf.net, "Orphans in the World" (UNICEF data, 2009), available at: www.numberof/netorphans-in-the-world (accessed July 15, 2011).

15. B. Klimes-Dugan and M. Gunnar, "Social Regulation of the Adrenocortical Response to Stress in Infants, Children, and Adolescents: Implications for Psychopathology and Education," in *Human Behavior, Learning, and the Developing Brain: Atypical Development*, ed. D. Coch, G. Dawson, and K. Fischer, New York: Guilford Press, 2007, 264–304.

16. A. Manne, *Motherhood*, Sydney: Allen and Unwin, 2005.

17. Ibid.

18. Ibid.

19. National Association of Child Care Resource and Referral Agencies, "Working Mothers Need Child Care," 2009, available at: http://www.naccrra.org/policy/background_issues/working_mothers.php (accessed July 1, 2009).

20. Manne, *Motherhood*.

21. D. L. Vandell, J. Belsky, M. Burchinal, L. Steinberg, and N. Vandergrift, "Do Effects of Early Child Care Extend to Age 15 Years? Results from the NICHD Study of Early Child Care and Youth Development," *Child Development* 81, no. 3 (2010): 737–756.

22. J. Belsky, "Emanuel Miller Lecture: Developmental Risks (Still) Associated with Early Child Care," *Journal of Child Psychology and Psychiatry* 42, no. 7 (October 2001): 845–859.

23. Manne, *Motherhood*.

24. M. Bunting, "Nursery Tales," *The Guardian*, July 8, 2004, available at: http://www.guardian.co.uk/education/2004/jul/08/schools.uk (accessed July 2, 2009).

25. Rebecca Adams quote in Manne, *Motherhood*, 230.

26. Manne, *Motherhood*.

27. E. S. Peisner-Feinberg, M. Burchinal, R. M. Clifford, et al., *The Children of the Cost, Quality, and Outcomes Study Go to School,* Chapel Hill: University of North Carolina Press, 1999; K. R. Thornburg, W. A. Mayfield, J. S. Hawks, and K. L. Fuger, *The Missouri Quality Rating System School Readiness Study: Executive Summary,* Columbia: University of Missouri, the Center for Family Policy and Research and the Institute for Human Development, 2009.

28. Thornburg et al., *The Missouri Quality Rating System School Readiness Study.*

29. Peisner-Feinberg et al., *The Children of the Cost, Quality, and Outcomes Study Go to School.*

30. National Association of Child Care Resource and Referral Agencies, "Working Mothers Need Child Care."

Chapter 6

1. A. Timm, "A Family's Story: Perinatal Depression," Voices for Illinois Children, March 2008, available at: http://www.voices4kids.org/issues/files/perinatdep.pdf (accessed January 2008).

2. E. Z. Tronick, "Infant Moods and the Chronicity of Depressive Symptoms: The Co-creation of Unique Ways of Being Together for Good or Ill," paper 1, "The Normal Process of Development and the Formation of Moods," *Journal of Psychosomatic Medicine and Psychotherapy* 49, no. 4 (2003): 408–424.

3. T. Field, D. B. Estroff, R. Yando, C. del Valle, J. Malphurs, and S. Hart, "'Depressed' Mothers' Perceptions of Infant Vulnerability Are Related to Later Development," *Child Psychiatry and Human Development* 27, no. 1 (Fall 1996): 43–53.

4. T. Field, "Infants of Depressed Mothers," *Infant Behavior and Development* 18, no. 1 (1995): 1–13.

5. F. Thomson Salo, "The Trauma of Depression in Infants: A Link with Attention Deficit Hyperactivity Disorder?" in *Cries Unheard: A New Look at Attention Deficit Hyperactivity Disorder,* ed. G. Halasz, G. Anaf, P. Ellingsen, A. Manne, and F. Thomson Salo, Melbourne: The Learner, 2003, 61–73.

6. R. Pianta, B. Egeland, and M. Farrell-Erickson, "The Antecedents of Maltreatment: Results of the Mother-Child Interaction Research Project," in *Child Maltreatment: Theory and Research on the Causes and Consequences of Child Abuse and Neglect,* ed. D. Cicchetti and V. Carlson, New York: Cambridge University Press, 1989, 203–253.

7. A. F. Lieberman and K. Knorr, "The Impact of Trauma: A Developmental Framework for Infancy and Early Childhood," *Pediatric Annals* 36, no. 4 (April 2007): 209–215.

8. R. Pianta, B. Egeland, and M. Farrell-Erickson, "The Antecedents of Maltreatment: Results of the Mother-Child Interaction Research Project," in Cicchetti and Carlson, *Child Maltreatment,* 203–253.

9. Thomson Salo, "The Trauma of Depression in Infants."

10. Every Child Matters Education Fund, *We Can Do Better: Child Abuse and Neglect Deaths in America,* Washington, D.C., September 2010.

11. Lieberman and Knorr, "The Impact of Trauma."

12. Ibid.

13. B. Perry and M. Szalavitz, *The Boy Who Was Raised as a Dog, and Other Stories from a Child Psychiatrist's Notebook*, New York: Basic Books, 2006.

14. Ibid.

15. B. Perry, "Violence and Childhood: How Persisting Fear Can Alter the Developing Child's Brain," 2001, available at: http://www.terrylarimore.com/PainAndViolence.html.

16. M. Ichise, D. C. Vines, T. Gura, et al., "Effects of Early Life Stress on [11C]DASB Positron Emission Tomography Imaging of Serotonin Transporters in Adolescent Peer- and Mother-Reared Rhesus Monkeys," *Journal of Neuroscience* 26, no. 17 (April 26, 2006): 4638–4643.

17. Ibid.

18. D. English, D. Marshall, and A. Stewart, "Effects of Family Violence on Child Behavior and Health During Early Childhood," *Journal of Family Violence* 18, no. 1 (2003): 43–57.

19. A. Danese, C. M. Pariante, A. Caspi, A. Taylor, and R. Poulton, "Childhood Maltreatment Predicts Adult Inflammation in a Life-Course Study," *Proceedings of the National Academy of Sciences USA* 104, no. 4 (2007): 1319–1324.

20. S. E. Taylor, J. S. Lerner, R. M. Sage, B. J. Lehman, and T. E. Seeman, "Early Environment, Emotions, Responses to Stress, and Health," *Journal of Personality* 72, no. 6 (December 2004): 1365–1393.

21. Ibid.

22. Ibid.

23. M. H. Teicher, S. L. Andersen, A. Polcari, C. M. Anderson, and C. P. Navalta, "Developmental Neurobiology of Childhood Stress and Trauma," *Psychiatric Clinics of North America* 25, no. 2 (June 2002): vii–viii, 397–426.

24. B. S. McEwen and E. Norton, *The End of Stress as We Know It*, Washington, D.C.: National Academies Press, 2002.

25. Teicher et al., "Developmental Neurobiology of Childhood Stress and Trauma."

26. M. H. Teicher, J. A. Samson, A. Polcari, and C. E. McGreenery, "Sticks, Stones, and Hurtful Words: Relative Effects of Various Forms of Childhood Maltreatment," *American Journal of Psychiatry* 163, no. 6 (June 2006): 993–1000.

27. Ibid.

28. S. A. Graham-Bermann and J. Seng, "Violence Exposure and Traumatic Stress Symptoms as Additional Predictors of Health Problems in High-Risk Children," *Journal of Pediatrics* 146, no. 3 (March 2005): 349–354.

Chapter 7

1. B. Perry, "Childhood Experience and the Expression of Genetic Potential: What Childhood Neglect Tells Us About Nature and Nurture," *Brain and Mind* 3 (2002): 79–100.

2. P. R. Levitt, "Nature and Nurture: New Techniques and Insights Reveal the Amazing Complexity of the Human Brain," *Lens* (2003), available at: http://www.mc.vanderbilt.edu/lens/article/?id=69 (accessed March 4, 2011).

3. P. Gluckman, *Mismatch: Why Our World No Longer Fits Our Bodies*, Oxford: Oxford University Press, 2006.

4. J. Qiu, "Epigenetics: Unfinished Symphony," *Nature International Weekly Journal of Science* 441 (2006): 143–145.

5. Gluckman, *Mismatch.*

6. For an excellent exploration of the topic of epigenetics, see *Ghost in Your Genes*, a 2006 BBC documentary produced and directed by Sarah Holt and Nigel Patterson, broadcast in 2007 on the PBS series *Nova*, and available at: http://www.pbs.org/wgbh/nova/genes/.

7. Ibid.

8. R. Karr-Morse and M. Wiley, *Ghosts from the Nursery: Tracing the Roots of Violence*, New York: Atlantic Monthly Press, 1997.

9. E. Watters, "DNA Is Not Destiny," *Discover*, November 22, 2006, available at: http://discovermagazine.com/2006/nov/cover.

10. S. Begley, "The Sins of the Father: Take 2," *Newsweek*, January 17, 2009; R. Highfield, "I Blame My Grandparents," *The Telegraph*, January 10, 2006.

11. Begley, "The Sins of the Father."

12. Holt and Paterson, *Ghost in Your Genes.*

13. Ibid.

14. G. Kolata, "So Big and Healthy Grandpa Wouldn't Even Know You," *New York Times*, July 30, 2006.

15. M. D. Anway, C. Leathers, and M. K. Skinner, "Endocrine Disruptor Vinclozolin Induced Epigenetic Transgenerational Adult-Onset Disease," *Endocrinology* 147, no. 12 (December 2006): 5515–5523.

16. Ibid.

17. L. A. Pray, "Epigenetics: Genome, Meet Your Environment," *The Scientist* 18, nos. 13–14 (July 5, 2004).

18. J. G. Falls, D. J. Pulford, A. A. Wylie, and R. L. Jirtle, "Genomic Imprinting: Implications for Human Disease," *American Journal of Pathology* 154, no. 3 (March 1999): 635–647.

19. C. Brownlee, "Nurture Takes the Spotlight: Decoding the Environment's Role in Development and Disease," *Science News*, June 24, 2006, available at: http://www .thefreelibrary.com/Nurture_takes_the_spotlight:_decoding_the_environment's_role in_._._._-a0148858116 (accessed March 11, 2010).

20. J. Kagan, "Biology, Context, and Developmental Inquiry," *Annual Review of Psychology* 54 (online: June 10, 2002): 1–23.

21. Holt and Paterson, *Ghost in Your Genes.*

22. Watters, "DNA Is Not Destiny."

23. R. A. Simmons, "Developmental Origins of Diabetes: The Role of Epigenetic Mechanisms," *Current Opinion in Endocrinology, Diabetes, and Obesity* 14, no. 1 (February 2007): 13–16.

24. L. M. Villeneuve and R. Natarajan, "The Role of Epigenetics in the Pathology of Diabetic Complications," *American Journal of Physiology—Renal Physiology* 299, no. 1 (July 2010): F14–F25.

25. H. Mano, "Epigenetic Abnormalities in Cardiac Hypertrophy and Heart Failure," *Environmental Health and Preventive Medicine* 13, no. 1 (January 2008): 25–29.

26. B. Weinhold, "Epigenetics: The Science of Change," *Environmental Health Perspectives* 114, no. 3 (March 2006): A160–A167.

27. B. McVittie, "Abuse Affects Genes," February 2009, available at: http://epigenome .eu/en/1,65,0 (accessed March 10, 2010).

28. M. D. Klinnert, H. S. Nelson, M. R. Price, A. D. Adinoff, D. Y. Leung, and D. A. Mrazek, "Onset and Persistence of Childhood Asthma: Predictors from Infancy," *Pediatrics* 108, no. 4 (October 2001): E69.

29. Genetic Science Learning Center, "The New Science of Addiction: Genetics and the Brain," 2005, available at: http://teach.genetics.utah.edu/content/addiction/webquest/ Exploring%20The%20New%20Science%20of%20Addiction.pdf (accessed March 10, 2010).

30. Ibid.

31. Ibid.

32. C. Wallis, "The Genetics of Addiction," *Fortune*, October 16, 2009, http://money .cnn.com/2009/10/16/news/genes_addiction.fortune/index.htm (accessed March 10, 2010).

33. Ibid.

34. Ibid.

35. J. I. Nurnberger and L. J. Bierut, "Seeking the Connections: Alcoholism and Our Genes," *Scientific American* 296, no. 4 (2007): 46–53.

36. S. Borenstein, "Smoking, Lung Cancer, May Be in Genes," *The Oregonian*, April 3, 2008.

37. C. Gazit, *This Emotional Life*, NOVA/WGBH Science Unit and Vulcan Productions, January 5, 2010.

38. N. Risch, R. Herrel, T. Lehner, et al., "Interaction Between the Serotonin Transporter Gene (5-HTTLPR), Stressful Life Events, and Risk of Depression: A Meta-Analysis," *Journal of the American Medical Association* 301, no. 23 (June 17, 2009): 2462–2471.

39. Ibid.

40. National Institute of Mental Health, "Gene Variants Protect Against Adult Depression Triggered by Childhood Stress," February 4, 2008, available at: http://www.nimh.nih .gov/science-news/2008/gene-variants-protect-against-adult-depression-triggered-by-child hood-stress.shtml (accessed March 10, 2010).

41. L. Edwards, "Early Life Stress Has Effects at the Molecular Level," *Nature Neuroscience*, November 12, 2009, available at: http://www.physorg.com/news177227567.html.

42. "Connection Between Depression and Osteoporosis Shown by Hebrew University Researchers," *ScienceDaily*, October 31, 2006, available at: http://www.sciencedaily .com/releases/2006/10/061030183243.htm.

43. J. Hamilton, "Cyber Scout Puts Autism Studies on Faster Track," NPR, April 9, 2009, available at: http://www.npr.org/templates/story/story.php?storyId=102852254 (accessed March 10, 2010).

44. K. Doheny, "Depression and Alzheimer's Linked," WebMD, April 7, 2008, available at: http://www.webmd.com/depression/news/20080407/depression-and-alzheimers -linked (accessed March 10, 2010).

45. E. B. Binder, R. G. Bradley, W. Liu, et al., "Association of FKBP5 Polymorphisms and Childhood Abuse with Risk of Posttraumatic Stress Disorder Symptoms in Adults," *Journal of the American Medical Association* 299, no. 11 (March 19, 2008): 1291–1305.

46. Casey Family Programs, *Assessing the Effects of Foster Care: Mental Health Outcomes from the Casey National Alumni Study*, 2010.

47. Child Welfare Information Gateway, "Foster Care Statistics 2009," *Numbers and Trends*, 2011, available at: http://www.childwelfare.gov/pubs/factsheets/foster.cfm#key (accessed June 19, 2010).

48. Falls et al., "Genomic Imprinting."

49. G. Oh and A. Petronis, "Environmental Studies of Schizophrenia Through the Prism of Epigenetics," *Schizophrenia Bulletin* 34, no. 6 (November 2008): 1122–1129.

50. S. Boyles, "Schizophrenia, Bipolar Disorder: Gene Link?" WebMD, January 15, 2009, available at: http://www.webmd.com/bipolar-disorder/news/20090115/schizophrenia-bipolar-disorder-gene-link (accessed March 10, 2010).

51. D. Steinberg, "Determining Nature vs. Nurture," *Scientific American Mind* 17, no. 5 (2006): 12–14.

52. J. Hamilton, "Schizophrenia May Be Linked to Immune System," NPR, July 1, 2009, available at: http://www.npr.org/templates/story/story.php?storyId=106151437 (accessed March 10, 2010).

53. P. Tienari, L. C. Wynne, A. Sorri, et al., "Genotype-Environment Interaction in Schizophrenia-Spectrum Disorder: Long-Term Follow-Up Study of Finnish Adoptees," *British Journal of Psychiatry* 184 (March 2004): 216–222.

54. For additional information, see Epigenetics? (http://epigenome.eu/en/4,27,0) and University of Utah Genetic Science Learning Center (http://learn.genetics.utah.edu/).

55. Alzheimer's Association, "Key Facts About Alzheimer's Disease," *Alzheimer's Disease Facts and Figures*, 2010, available at: http://www.alz.org/alzheimers_disease_facts_figures.asp#key (accessed May 5, 2010).

56. N. H. Zawia, D. K. Lahiri, and F. Cardozo-Pelaez, "Epigenetics, Oxidative Stress, and Alzheimer Disease," *Free Radical Biology and Medicine* 46, no. 9 (May 1, 2009): 1241–1249.

57. C. Johnson, "Autism Rates: Government Studies Find 1 in 100 Children Has Autism Disorders," October 5, 2009, available at: http://www.huffingtonpost.com/2009/10/05/autism-rates-government-s_n_309290.html (accessed March 10, 2010). See also Centers for Disease Control and Prevention, "Autism Spectrum Disorders (ASDs): Data and Statistics," available at: http://www.cdc.gov/ncbddd/autism/data.html.

58. D. K. Kinney, A. M. Miller, D. J. Crowley, E. Huang, and E. Gerber, "Autism Prevalence Following Prenatal Exposure to Hurricanes and Tropical Storms in Louisiana," *Journal of Autism and Developmental Disorders* 38, no. 3 (March 2008): 481–488.

59. "New Theory of Autism Suggests Symptoms or Disorder May Be Reversible," *ScienceDaily*, April 2, 2009, available at: http://www.sciencedaily.com/releases/2009/04/090401145312.htm (accessed March 10, 2010).

60. National Institutes of Health and National Institute of Environmental Health Sciences, "NIH to Fund Grants for Epigenetic Research," September 17, 2009, available at: http://www.news-medical.net/news/20090917/NIH-to-fund-grants-for-epigenetic-research.aspx (accessed March 11, 2010).

61. U.S. Department of Health and Human Services, National Institutes of Health, "NIH Funds Grantees Focusing on Epigenomics of Human Health and Disease," *NIH News*, September 16, 2009, available at: http://www.nih.gov/news/health/sep2009/od-16.htm (accessed March 10, 2010).

62. Oh and Petronis, "Environmental Studies of Schizophrenia Through the Prism of Epigenetics."

Chapter 8

1. A. Manne, *Motherhood*, Sydney: Allen and Unwin, 2005.

2. S. W. Porges, "The Polyvagal Theory: Phylogenetic Substrates of a Social Nervous System," *International Journal of Psychophysiology* 42, no. 2 (October 2001): 123–146.

3. S. W. Porges, "Social Engagement and Attachment: A Phylogenetic Perspective," *Annals of the New York Academy of Sciences* 1008 (December 2003): 31–47.

4. S. Aslanian, "Researchers Still Learning from Romania's Orphans," NPR, *All Things Considered*, September 16, 2006, available at: http://www.npr.org/templates/story/story.php?storyId=6089477.

5. R. Dykema, "How Your Nervous System Sabotages Your Ability to Relate: An Interview with Stephen Porges About His Polyvagal Theory," *NEXUS: Colorado's Holistic Journal*, (March–April 2006), available at: http://www.nexuspub.com/articles_2006/interview_porges_06_ma.php.

6. See the website for the Unicorn Children's Foundation, www.unicornchildrensfoundation.org (accessed June 19, 2010).

7. D. Siegel, *The Developing Mind: How Relationships and the Brain Interact to Shape Who We Are*, New York: Guilford Press, 1999, 337.

8. NICHD Early Child Care Research Network, "Infant-Mother Attachment Classification: Risk and Protection in Relation to Changing Maternal Caregiving Quality," *Developmental Psychology* 42, no. 1 (January 2006): 38–58; R. Schaffer, "Forming Relationships," in *Introducing Child Psychology*, Oxford: Blackwell, 2004, 83–121.

9. A. N. Schore, "Back to Basics: Attachment, Affect Regulation, and the Developing Right Brain: Linking Developmental Neuroscience to Pediatrics," *Pediatric Review* 26, no. 6 (June 2005): 204–217.

10. A. N. Schore, "The Effects of a Secure Attachment Relationship on Right Brain Development, Affect Regulation and Infant Mental Health," *Infant Mental Health Journal* 22 (2001): 7–66.

11. Ibid.

12. A. N. Schore, "Dysregulation of the Right Brain: A Fundamental Mechanism of Traumatic Attachment and the Psychopathogenesis of Posttraumatic Stress Disorder," *Australian and New Zealand Journal of Psychiatry* 36, no. 1 (February 2002): 9–30.

13. G. Maté, *When the Body Says No: Understanding the Stress-Disease Connection*, New York: John Wiley, 2003.

14. A. Goodman, "In the Realm of Hungry Ghosts: Interview with Dr. Gabor Maté," *Democracy Now!* February 3, 2010, available at: http://www.democracynow.org/2010/2/3/addiction (accessed April 28, 2010).

15. A. N. Schore, "Attachment Trauma and the Developing Right Brain: Origins of Pathological Dissociation," in *Dissociation and the Dissociative Disorders: DSM-V and Beyond*, ed. P. Dell and J. O'Neil, New York: Routledge, 2009, 107–141.

Chapter 9

1. M. Sykes-Wylie, "Beyond Talk: Using Our Bodies to Get to the Heart of the Matter," *Psychotherapy Networker* (July–August 2004).

2. Ibid.

3. See the website of the EMDR Institute, www.emdr.com (accessed June 24, 2010).

4. For an in-depth look at state-of-the-art trauma therapy for children, visit Dr. Bruce Perry's Child Trauma Academy in Houston, Texas (www.childtraumaacademy.org); see also Bruce Perry and Maia Szalavitz, *The Boy Who Was Raised as a Dog, and Other Stories from a Child Psychiatrist's Notebook—What Traumatized Children Can Teach Us About Loss, Love, and Healing*, New York: Basic Books, 2007.

5. EMDR Institute, www.emdr.com (accessed June 24, 2010).

6. See myShrink, the website of Dr. Suzanne LaCombe, http://www.myshrink.com (accessed June 19, 2010).

7. P. Ogden and K. Minton, "One Method for Processing Traumatic Memory," *Traumatology* 6, no. 3 (2000).

8. D. Grand, "What Is Brainspotting?" available at: http://www.brainspotting.pro/page/what-brainspotting (accessed June 19, 2010).

9. See www.eft-therapy.com or www.emofree.com.

10. S. G. Benish, Z. E. Imel, and B. E. Wampold, "The Relative Efficacy of Bona Fide Psychotherapies for Treating Post-Traumatic Stress Disorder: A Meta-analysis of Direct Comparisons," *Clinical Psychology Review* 28, no. 5 (June 2008): 746–758.

11. M. Hubble, B. Duncan, and S. Miller, eds., *The Heart and Soul of Change: What Works in Therapy*, Washington, D.C.: American Psychological Association, 1999.

12. D. Siegel, *The Developing Mind: How Relationships and the Brain Interact to Shape Who We Are*, New York: Guilford Press, 1999.

13. J. R. Schore and A. N. Schore, "Modern Attachment Theory: The Central Role of Affect Regulation in Development and Treatment," *Journal of Social Work* 36 (2008): 9–20.

14. Boston University, "Yoga May Elevate Brain GABA Levels, Suggesting Possible Treatment for Depression," *ScienceDaily*, May 22, 2007.

15. M. R. Irwin, R. Olmstead, and M. N. Oxman, "Augmenting Immune Responses to Varicella Zoster Virus in Older Adults: A Randomized, Controlled Trial of Tai Chi," *Journal of the American Geriatric Society* 55, no. 4 (April 2007): 511–517.

16. M. Kaufman, "Meditation Gives Brain a Charge, Study Finds," *Washington Post*, January 3, 2005.

17. Ibid.

18. Z. V. Segal, J. M. G. Williams, and J. D. Teasdale, "Mindfulness-Based Cognitive Therapy for Depression: A New Approach to Preventing Relapse," in *Mindfulness and the Therapeutic Relationship*, ed. S. F. Hick and T. Bien, New York: Guilford Press, 2008, 351; M. Linehan, *Cognitive-Behavioral Treatment of Borderline Personality Disorder*, New York: Guilford Press, 1993; K. Witkiewitz and G. Marlatt, eds., *Therapist's Guide to Evidence-Based Relapse Prevention*, Burlington, Mass.: Academic Press, 2007; J. Kabat-Zinn, "Mindfulness-Based Interventions in Context: Past, Present, and Future," *Clinical Psychology: Science and Practice* 10, no. 2 (2006): 144–156.

19. Kabat-Zinn, "Mindfulness-Based Interventions in Context."

20. R. W. Sears, "Mindfulness in Clinical Practice: A Basic Overview," *Ohio Psychologist* 56 (August 2009).

21. D. Siegel, *The Mindful Brain: Reflection and Attunement in the Cultivation of Well-being*, New York: W. W. Norton, 2007.

22. Z. V. Segal, J. D. Teasdale, and J. M. G. Williams, "Mindfulness-Based Cognitive Therapy: Theoretical Rationale and Empirical Status," in *Mindfulness and Acceptance*, ed. S. C. Hayes, V. M. Follete, and M. Linehan, New York: Guilford Press, 2004.

23. Siegel, *The Mindful Brain*, xiv.

24. E. Sternberg, *The Balance Within: The Science Connecting Health and Emotions*, New York: W. H. Freeman, 2001.

25. University of Minnesota, "Taking Charge of Your Health: Interview with Esther Sternberg," available at: http://www.takingcharge.csh.umn.edu/interviews/interview-esther -sternberg-0.

26. See http://thinkexist.com/quotation/i_believe_that_the_very_purpose_of_life_is_to _be/145371.html.

Chapter 10

1. A. Goodman, "In the Realm of Hungry Ghosts: Interview with Dr. Gabor Maté," *Democracy Now!* February 3, 2010, available at: http://www.democracynow.org/2010/ 2/3/addiction (accessed April 28, 2010).

2. U.S. Department of Health and Human Services, Administration for Children and Families, "Child Abuse and Neglect Statistics, " available at: http://www.childwelfare.gov/ systemwide/statistics/can.cfm (accessed June 23, 2010).

3. Fight Crime: Invest in Kids, "A Cycle of Child Abuse, Crime, and Violence," available at: http://www.fightcrime.org/issues/child-abuse-neglect (accessed March 16, 2010).

4. Centers for Disease Control and Prevention, "HIV/AIDS Surveillance Report: Diagnoses of HIV Infection and AIDS in the United States and Dependent Areas, 2009," available at: http://www.cdc.gov/hiv/topics/surveillance/basic.htm (accessed June 23, 2010); Centers for Disease Control and Prevention, "Heart Disease Facts," available at: http://www.cdc.gov/heartdisease/facts.htm (accessed June 23, 2010).

5. United Nations Children's Fund (UNICEF), *The State of the World's Children: 2010 Child Rights*, New York: UNICEF, November 2009.

6. Ibid.

7. Ibid.

8. United Nations Children's Fund (UNICEF), *Facts on Children*, New York: UNICEF, April 2007.

9. Ibid.

10. Ibid.

11. V. Attanayake, R. McKay, M. Joffres, S. Singh, F. Burkle Jr., and E. Mills, "Prevalence of Mental Disorders Among Children Exposed to War: A Systematic Review of 7,920 Children," *Medicine, Conflict, and Survival* 25, no. 1 (January–March 2009): 4–19.

12. S. al-Haideri, "Experts Fear a Lost Generation," *Iraqi Crisis Reports* 237 (June 11, 2007), available at: http://iwpr.net/report-news/experts-fear-lost-generation.

13. Ibid.

14. UNICEF, *Facts on Children*, available at: http://www.unicef.org/media/media _fastfacts.html (accessed June 6, 2011).

Appendix B

1. Information in appendix B is drawn from S. J. Dallam, "The Long-Term Medical Consequences of Childhood Trauma," in *The Cost of Child Maltreatment: Who Pays? We All Do*, ed. K. Franey, R. Geffner, and R. Falconer, San Diego: Family Violence and Sexual Assault Institute, 2001, 1–14.

INDEX